# World Wisdom
# The Library of Perennial Philosophy

The Library of Perennial Philosophy is dedicated to the exposition of the timeless Truth underlying the diverse religions. This Truth, often referred to as the *Sophia Perennis*—or Perennial Wisdom—finds its expression in the revealed Scriptures as well as the writings of the great sages and the artistic creations of the traditional worlds.

*Philosophy of Science in the Light of the Perennial Wisdom* appears as one of our selections in the Perennial Philosophy series.

# The Perennial Philosophy Series

In the beginning of the twentieth century, a school of thought arose which has focused on the enunciation and explanation of the Perennial Philosophy. Deeply rooted in the sense of the sacred, the writings of its leading exponents establish an indispensable foundation for understanding the timeless Truth and spiritual practices which live in the heart of all religions. Some of these titles are companion volumes to the Treasures of the World's Religions series, which allows a comparison of the writings of the great sages of the past with the perennialist authors of our time.

# Philosophy of Science in the Light of the Perennial Wisdom

Mahmoud Bina
&
Alireza K. Ziarani

World Wisdom

Library of Congress Cataloging-in-Publication Data
Names: Bina, Mahmoud, 1938- author. | Ziarani, Alireza K., 1972- author.
Title: Philosophy of science in the light of the perennial wisdom / by
Mahmoud Bina & Alireza K. Ziarani.
Description: Bloomington, Indiana : World Wisdom, 2020. | Includes
bibliographical references and index. | Summary: "Backed by its
technological achievements, modern science appears as the de facto
source of truth to the majority of our contemporaries. Its sole reliance
on reason and empirical data gives it an air of objectivity that has
conferred upon it an almost unquestioning authority. Against the
backdrop of this pervasive scientism, Philosophy of Science in the Light
of the Perennial Wisdom is a daring attempt to offer an intellectual
critique of modern science in its foundation by rigorously examining the
intrinsic limitations of rational thought and empirical investigation.
Unique of its kind, this book offers a refreshing look at the
traditional doctrines of epistemology and metaphysics as an antidote to
the subjective as well as objective errors of modern science, which is
thus revealed as no more than a belief system that falls radically short
of offering a full knowledge of reality; this, in contrast to the
perennial wisdom of the world's great religions that for millennia have
offered humankind not only keys to true knowledge, but also the means of
attaining it, which precisely constitutes man's reason for being"--
Provided by publisher.
Identifiers: LCCN 2020027468 (print) | LCCN 2020027469 (ebook) |
ISBN 9781936597680 (hardback) | ISBN 9781936597697 (epub)
Subjects: LCSH: Knowledge, Theory of. | Science--Philosophy. |
Wisdom. | Metaphysics.
Classification: LCC BD181 .B48 2020 (print) | LCC BD181 (ebook) |
DDC 121--dc23
LC record available at https://lccn.loc.gov/2020027468
LC ebook record available at https://lccn.loc.gov/2020027469

Printed on acid-free paper in the United States of America

For information address World Wisdom, Inc.
P.O. Box 2682, Bloomington, Indiana 47402-2682
www.worldwisdom.com

# CONTENTS

# AUTHORS' NOTE

All quotes from the Bible are from the King James Version. The quoted passages from Plato's dialogues are translated from Greek in consultation with their English translations by Benjamin Jowett and R.G. Bury and their French translations by Albert Rivaud and Victor Cousin. The quoted Koranic verses are translated from Arabic in consultation with their English translations by Marmaduke Pickthall and A. Yusuf Ali. The quoted poems from Rumi are translated from Persian in consultation with their English translation by R.A. Nicholson. All other translations of the quoted Arabic and Persian texts are by the authors. Consistent with the traditional usage of terms, modern references to gender are not addressed, and the pronoun "he" and the term "man" are used to refer to any human being.

# PREFACE

This is not yet another academic book on the philosophy of science. Unlike the numerous textbooks and research monographs that exist on the subject, this book is a critique of modern science, its premises and its claims, and an account of an alternative perspective on science based on the wisdom of the divinely instituted traditions that have fashioned humanity since time immemorial. As alien as this alternative perspective may appear to the modern mentality, this was how mankind perceived science throughout all the ages except for the past few centuries. It is all-too-convenient for the modern man of scientistic mind to criticize the worldview of the ancients and their approach to science. Our ancestors are constantly put on trial *in absentia* and their thoughts are often trivialized by the moderns, with little chance, if any, of being properly heard.

How, then, did the ancients inquire about knowledge and reality? How would they view our modern science if they had the opportunity? How would they respond to the modern criticism of their outlook? This book aims to answer these questions. Far from appealing to religious sentiments on a superficial level, we intend to offer intellectual arguments in support of the views of the great traditions on knowledge and reality. Modern science, with all the wonders it works and all the catastrophes it may be complicit in, is assessed here not in its results, beneficial or detrimental as they may be, but in its principles vis-à-vis the wisdom of the great traditions, which have offered humanity countless men of wisdom and greatness nearly everywhere and throughout almost all the ages.

The modern academic approach to scholarship is cut to the measure of the modern man of scientific mind. Consistent with the central thesis of this book, we do not endorse this approach, and, as such, we do not find ourselves obliged to adhere to its academic conventions, whence the non-conventional character of an exposition that, we hope, the reader will soon find himself immersed in.

# INTRODUCTION

"I think, therefore I am," wrote Descartes, and thus heralded the Age of Reason—an age in which man wished to free science of all prejudice by grounding it in "pure reason." Countless other "thinkers" followed him, right up to the present age, each offering a new "philosophy," at times contradicting the preceding ones, at times corroborating them, at times elaborating on them, and at times opening new chapters of "thought"; each, of course, expecting others to take their thoughts as expressions of truth, even those who questioned the very notion of truth itself.

But how do we know if anything is true? Some consider logic—the science of reasoning—as the fountain of truth: Descartes' "therefore" signals a proof, hence rational thought. But how can we be sure that the rules of logic are valid, that our premises are well founded, and that logic can give us a full account of truth? Others look to sense experience for the foundation of truth: only that is true which is empirically verifiable—or alternatively, logically or empirically falsifiable. But how, then, do we know that experience is the criterion of truth? Yet others see in human sentiment the criterion of truth: that is true which takes account of our feelings as individual sentient beings and, for example, gives us an immediate feeling of happiness. And again, we will ask, what is it that guarantees the truth of such an idea?

But before we can begin to look for the criterion of truth, we have to ascertain that there is such a thing as truth. Now, unless one accepts that there is indeed such a thing as truth, nothing holds: remove truth, and everything collapses. Nothing can get around this: "there is no truth" cannot be taken as true since it would *ipso facto* refute itself. Why is it that self-refutation, inconsistency in logical terms, matters? Because an inconsistent thesis, that is to say, a thesis that bears its refutation within itself, has no real persuasive power: to expect others to believe in it would be to expect them to take it as true, hence to expect them not to believe in it. In that case, one would have to keep it to oneself and say: "I do not know." That does not mean, however, that "no one else knows, either," since that would imply that one knows that the proposition "no one else knows" is true, which would contradict the initial "I do not know." In fact, even to say "I do not know"—in response to the question of whether one knows if there is such a thing as truth—would be inconsistent, for to say "I do not know" is to imply that I know that one thing is true and that is precisely the fact that "I do not know." He who does not take the self-evidence of truth for granted is bound to remain silent, or else ends up contradicting himself.

The denial of truth, therefore, or doubt about it, cannot be the starting point of any logical system of thought: to avoid inconsistency—that is, to remain logical—one has to assume that there is such a thing as truth before one presents any reasoning. This is a fundamental "dogma" without which nothing holds. It shows, moreover, that no system of thought can be without a dogma. A "dogma-free" starting point is itself a dogma, though a self-contradictory one.

To say that there *is* such a *thing* as *truth* is to speak in *absolute* terms. The notions of "being," "reality,"[1] "truth," and "absoluteness" are thus intrinsically interrelated. Reason cannot prove them; on the contrary, it takes them for granted, that is to say, it cannot function in their absence.

Anyone who has any judgment about anything and communicates it to others has already assumed that what he tells them will mean essentially the same thing to them, and that they will recognize the truth of his opinion, that is, they will have the same judgment. If what one man says has a mean-

---

1 The origin of the English word "reality" is the Latin *res*, meaning "thing."

ing that is accessible solely to him, and is true solely for him, why then would he even say it to others? In other words, he has already assumed (i) that there is such a thing as truth, (ii) that this truth corresponds to a reality, and (iii) that he, as well as others, has access to this reality. In other words, he has assumed that the truth of what he says relates to an *objective* reality that is independent of the human *subject* who says it or hears it. He has implicitly accepted the notion of *objectivity*. Unless one accepts that man is fundamentally objective, one quickly finds oneself in refutation of oneself.

One has to start with the self-evidence of objective truth. Any attempt to deny the self-evidence of truth—or being, or reality, or absoluteness—will be self-defeating. All one can apparently do to escape this is to be silent and let others speak. This, however, is precisely the opposite of what the man of "reason" has so often done in the course of the past few centuries—he who wished to ground everything in reason. Never before, in the history of humanity, did man think, speculate, philosophize, and speak as much. Yet never before was he more inconsistent: if the man of the Age of Reason had been reasonable about reason, he would have taken note of its inherent limitations;[2] as such, he would not have taken reason as the sole agent of the discovery of truth, and his science would not have been based solely on empirical and rational means; he would not have blamed the ancients for having had dogmas while thinking himself free of all prejudice, and so on and so forth. In short, he would not have done much of what he did over the course of the past few centuries.

It should be clear that in showing the inevitable inconsistency of the opinions that deny the notions of truth and objectivity, we are demonstrating the logical absurdity of

---

2 That is to say, first, that reason cannot function in a void; then, that it cannot, in and of itself, prove the truth of its methods; and finally, that it cannot guarantee that it can prove the truth of all that is true. Here we are presenting only a brief outline of the limitations of reason. An extensive treatment of this subject will follow in the next chapter.

many of the philosophical movements that have emerged since the dawn of the Age of Reason. Any system of thought that proposes an absolute principle while denying the notion of truth—hence the notion of objectivity—is condemned to self-refutation. This applies to subjectivism, in all its forms, which precisely denies the notions of truth and objectivity. The central thesis of every subjectivist opinion—various forms of the thesis that "there is no objective truth"—when applied to itself, refutes itself. Similarly, relativism carries the seed of its own negation within itself since it refutes its central thesis that "everything is relative" by this very thesis. Agnosticism, on the other hand, pretends to avoid the logical inconsistency of relativism by refraining from making a pronouncement on truth while implicitly promoting the idea that man is incapable of knowing it, that is, "I do not know" is taken as true of every man; in effect, therefore, it promotes the idea that "no one else knows, either." An agnostic, to be logical, would have to remain silent, in which case there would be no one to propose agnosticism. The arbitrary, hence self-refuting, nature of the philosophies that proclaim the primacy of experience or feeling is clear, for their overarching principles—which are mental formulations—can be neither experienced nor felt, precisely because mental formulations do not fall within the scope of experience or feeling. For example, the thesis that "experience is the criterion of truth" means that every truth will have to be empirically verifiable. Now, this thesis itself is not empirically verifiable. Consequently, the proposed thesis is not true according to its own measure. Empiricism, therefore, refutes itself. Not unrelated to this is any ideology that seeks to base truth on the experiences and feelings of man in his subjectivity and individuality: for example, various forms of existentialism, which place great emphasis on the individual's experiences and feelings.[3]

---

3 We shall return to some of these observations in more detail on various occasions throughout this book. Our aim here is to briefly show the

To repeat, we must accept that there is such a thing as truth; otherwise, we can proceed no further. But how would we recognize truth as truth when we are presented with it? Logic cannot be the sole source of truth since any logical system assumes the validity of certain rules of inference and takes as its starting point certain premises—axioms or postulates as well as primitive concepts. Yet it can prove neither the validity of its rules nor the truth of its premises. The truth of such premises can be acknowledged only if one possesses their essence deep within oneself. Thus, by virtue of having to assume the truth of certain premises in order to be able to function, reason itself points to its limited capacity as a means of the discovery of truth, and prompts us to look for the foundation of thought in something deeper within us than mere rational thought.[4]

Truth has its root in the notion of the Absolute. Without this ultimate point of reference, every logical argument is devoid of a foundation. Man cannot be certain of anything in the absence of this notion, because as soon as he becomes absolutely certain of anything without the implicit assumption of the notion of the Absolute, logically, he must let go of his certitude of it and start over again in a vicious circle of doubt. Moreover, only when man is certain can he truly find happiness; otherwise, his happiness would be devoid of a foundation, subject to fluctuations, and thus precarious.[5] The notion of the Absolute, the root of truth and certitude, without which no knowledge holds and no true happiness is possible, is what modern man has forgotten—the man who

degree to which the man of "reason" can be unreasonable. In his constant flight from the Absolute, the protagonist of reason—the rationalist—only proves himself unfaithful to reason.

4 "Reason destroys itself," says renowned mathematical physicist Roger Penrose (*New Scientist* [2008], Vol. 199, No. 2666, p. 49).

5 According to Virgil, "Happy is he who has been able to know the profound reasons of things" (*Georgics*, II.490). True happiness coincides with the knowledge of the truth.

thinks he "knows," and knows much better than traditional man,[6] that is, man before the Age of Reason, or man as he always was until a relatively recent phase in the history of mankind. In the absence of eternal notions such as the notion of the Absolute, which modern man is happy to leave behind, his knowledge—that is, modern science—is groundless. Traditional man, on the contrary, always based himself on the notion of the Absolute, and grounded everything, including his science, therein, whence the immense divergence between his conception of science and that of modern man.

It is befitting, therefore, in our search for the foundation of thought, and in seeking a philosophy of science, that we look to see how traditional man—man as he lived for the longest part of the history of mankind—viewed science, and how the perennial wisdom—the timeless truth underlying all the great traditions—regards the notions of knowledge and reality. If nothing else, our sense of justice, to say the least, should motivate us in this undertaking: *audiatur et altera pars*—"let the other side also be heard."

---

6 We use the terms "traditional man" and "the ancients" to refer to the state of humankind throughout history until, more or less, the end of the Middle Ages in Europe. "Modern man," on the other hand, is the term we use to refer to the state of humanity after the Renaissance, and especially after the Enlightenment. The demarcation line, not always clear, is when, in the thought of the so-called "thinkers" of the epoch, man as the image of God was practically replaced by man in his earthly contingency, that is, man cut off from his transcendent root.

## CHAPTER I

# FOUNDATIONAL QUESTIONS

### What is Science?

Few concepts have a stronger presence in the minds of our contemporaries than does science. But what is science? Can it be defined? When its meaning is not conditioned by a particular usage, science (from the Latin *scientia*) is synonymous with "knowledge"—and its lack, with ignorance. Science, or knowledge, cannot be defined, for in order to be able to define an object, one would first have to know what the object in question is, and thus, in order to be able to define knowledge, one would have to know it *a priori*. In other words, a definition for knowledge would imply prior knowledge of knowledge itself, the circular nature of any such definition showing its logical fallacy. Otherwise said, if knowledge is to be defined by means of itself, it will beg the question as to what knowledge is in the first place, and if knowledge is to be defined by means of something other than itself, then knowledge will have to be defined by means of ignorance (i.e. the lack of knowledge), which is absurd. This explains why science for traditional man was undefined,[1] similar to the no-

---

1 For example, in his *Text of Texts*, Haydar Amuli, a fourteenth-century Sufi philosopher, states that science as such cannot be defined, because a definition would have to be inclusive of all that lies within its scope and exclusive of all that lies outside its scope. Now, to be inclusive, it would have to include the Divine Knowledge, which in turn would imply that the knowledge of the human subject providing the definition would have to encompass the Divine Knowledge—which is impossible since the finite cannot encompass the Infinite. Science, therefore, is a premise like "being":

tions of point, line, and plane in geometry, the meanings of which we find within ourselves, and understand intuitively, but which we cannot define. One may be able to define such and such a branch of science by means of limiting the scope of the object of science, but science as such remains undefined and undefinable. This means that understanding its nature is not within the reach of pure reason. In other words, no rational criterion, that is to say, no formula, can be given to decide what science is or is not. Therefore, any attempt to propose a philosophy of science by mere rational means is self-refuting.

If science as such cannot be defined, what can we do to understand its meaning? One solution is to take note of those things that are called science in order to learn what is meant by science. For Plato, this solution is in fact the only one there is: Platonic *anamnesis* has it that learning is in fact recollection; in other words, if we did not bear knowledge in our immortal kernel, there would be no way for us to recognize it as such outside us. Thus, we will have to examine various things that are called science in order that we may intuit what science as such is. This methodology may be characterized as "phenomenological," not in the current philosophical usage of the term but in its etymological sense, implying careful study of the things referred to as science and their common features, with the aim of intuitively learning the underlying notion.

A point of caution is called for here. In our search for various types of science for our phenomenological examination, care must be taken not to limit the scope of science to its particular types such as the natural or mathematical sciences, as is so commonly done when it comes to modern conceptions

to be able to define "being," there has to *be* a subject who communicates the definition and another to whom it is communicated; there has to *be* a language within which the definition is given; and, moreover, there has to *be* a definition that can be given, and so on and so forth. These are all different ways of expressing fundamentally the same thesis, that is, that science, like the notions of truth and being, is undefinable.

of science. Such an arbitrarily imposed limitation is obviously unwarranted and is contrary to an objective inquiry, for any prior determination introduced by extrinsic criteria would imply a prior knowledge of knowledge as such—which, as noted above, would lead to self-refutation.

Accordingly, to present an understanding of the notion of science in the light of the wisdom of the great traditions, we shall present a phenomenological examination of science from their perspectives in some detail in a later chapter.[2] To put things into perspective, however, it is useful first to see how the modern philosophy of science views science.

## Modern Views of Science

The modern approach to the philosophy of science, such as Karl Popper's introduction of the criterion of falsifiability to distinguish science from everything else, often has only the *empirical* sciences in view: "A sentence (or a theory) is empirical-scientific if and only if it is falsifiable,"[3] he postulates, meaning that in principle it "must be capable of conflicting with possible, or conceivable, observations."[4] In other words, only that is scientific which allows for an experiment whereby, in principle, it could be proven false should such be the case.

It was Popper's dissatisfaction with the claim to scientific status of theories such as Freudian psychoanalysis or the Marxist theory of history that led him to his formulation of falsifiability as the criterion of (empirical) science. The un-

---

2 In chapter 5, "Science in the Mirror of the Religions."

3 Karl Popper, "Falsifizierbarkeit, zwei Bedeutungen von," in Helmut Seiffert and Gerard Radnitzky, *Handlexikon zur Wissenschaftstheorie*, second edition (Munich, Germany: Ehrenwirth Verlag, [1989] 1994), pp. 82-86. See also Karl Popper, *Logik der Forschung* (Vienna, Austria: Springer-Verlag, 1935), pp. 12-13.

4 Karl Popper, *Conjectures and Refutation: The Growth of Scientific Knowledge* (New York, NY: Basic Books, 1962), p. 39.

derlying thought in the falsificationist approach—in contrast to the inductivism of a Francis Bacon, for instance, for which science is the induction of laws from sets of data, or the verificationism of a John Locke, for example, which seeks to ground scientific theory in empirically-verifiable experience—is that while what is false can be proven as such either by logical argument or by experiment, what is capable of being proven false and is not confirmed as such by experiment is *likely* to be true, the certitude of its truth falling outside the reach of rational thought and the empirical investigation deriving from it.

The criterion of falsifiability for science is not falsifiable itself; as a result, it is not scientific by its own measure. The same is true of every other extrinsic criterion such as those underlying inductivism or empiricism: one cannot arrive at the principle of inductivism by induction nor can one empirically verify the principle of empiricism. Far from being objective and free of any "dogma," the modern conception of science is heavily based on extrinsic assumptions, which, viewed from an unbiased perspective, seem quite arbitrary.

Popper's falsificationism tries to avoid the logically unfounded claim of positivism, in its many forms, for which science consists only of what can be proven true either by logical argument or by experiment. Popper knew that logic itself would not support the positivist claim that logic or experience can prove the truth of everything that is true.[5] His principle of falsification, therefore, is deliberately formulated in such a way as to account only for the empirical sciences. Thus, while remaining a useful tool within its own limited scope, the principle of falsification cannot be the criterion of science as such.

Another criterion often invoked in this context as a necessary condition is repeatability or reproducibility. According

5 We made a brief reference to the limitations of logic in the Introduction, and will discuss it in detail in the next section.

to Popper, "Non-reproducible single occurrences are of no significance to science."[6] Why is it that reproducibility may be of interest? Because a thing that is experienced by someone, it is argued, will have to be capable of being experienced by others in order to have a universal and objective value. In other words, reproducibility gives us some measure of objectivity. This is a plausible criterion on its own plane, that is, within the domain of the objects of sensory perception. But it will not be applicable, at least not in its empirical sense, to the objects of man's faculties of knowledge that are beyond his sensory faculties—intuition, for instance: every man intuitively understands what a point is, but no man can empirically experience it as such.

In this context, it is worthwhile noting that not all that is pronounced science—sub-atomic physics, for instance—is within the reach of the common man. Repeating experiments of the kind often requires highly sophisticated laboratory environments and a high level of expertise that can only be achieved through years of study and research. Acceptance of the truth of the results of such experiments on the part of the common man of our time is a matter of belief and trust somewhat similar to the medieval man's acceptance of the truth of the spiritual experiences of the mystics who spoke of "the knowledge of God through experience" following, necessarily, a strict, and often highly-exacting, spiritual practice.[7] Admit-

6 Karl Popper, *The Logic of Scientific Discovery* (London, UK and New York, NY: Routledge, [1959] 2002), p. 66.

7 In one of his works about the saints of Seville whom he knew as a youth, Ibn Arabi recounts that he once asked one of them how he reached his station. "Close the door, disregard the intermediary means, keep the company of the Bounteous One, and He will speak to you without a veil," the saint replied. Ibn Arabi goes on to say that he followed the advice and achieved the state in question. It should be noted in passing that the ultimate goal of the spiritual man is seeking the truth and not a transitory experience. The juxtaposition of the experience of the modern scientist with that of the mystic, in principle ill-posed given the disparity of the nature and level of the two, is used here only to highlight the unjustifiability of the scientistic claim.

tedly, reproducibility in the context of the empirical sciences takes on a simpler expression than in those experiences which require more than mere sensory perception; as such, the criterion of reproducibility as applied to the empirical sciences cannot be expected to be directly applicable to all domains, and will have to be extended, and appropriately adapted, to be applicable to things that go beyond sensory experience. It should be clear that rejecting the experiences of the mystics reported throughout the ages, while accepting only those experiences which are discernible to the sensory faculties, is not based on any objective truth, but is an arbitrary exclusivism deriving from a self-refuting assumption—self-refuting, because one cannot empirically verify the overarching principle of empiricism itself.

Popper's theory was discussed here in some detail in order to illustrate the intrinsic contradiction of any philosophy that aims at defining science by means of some extrinsic criteria. One need not discuss each and every philosophy of science that has been proposed after Popper. As we noted, every rational criterion for defining science as such would lead to logical absurdity. As another example in this crowded field, one might mention Thomas Kuhn, for whom there is no science without subjective factors: there are only competing "paradigms," which are incommensurable with one another, and which shift over time.[8] Now, how is it that Kuhn is able to view the history of science from a vantage point that allows him to pronounce judgment on all science throughout all history? If taken seriously enough to be considered the basis of a new paradigm, his theory would give rise to a singularly objective paradigm that is universally true; since it supposedly takes account of all scientific paradigms throughout all history, it does not compete with other paradigms, is not incommensurable with them, and does not shift over time; it, there-

8 See Thomas S. Kuhn, *The Structure of Scientific Revolutions* (Chicago, IL: University of Chicago Press, [1962] 2012).

fore, is the very paradigm Kuhn declares inexistent. Like all subjectivism, Kuhn's "revolutionary" theory carries the seed of its own negation within itself.

## The Limitations of Reason

Our preceding arguments demonstrate the logical absurdity of any attempt to ground truth and science solely in reason, indicating that reason is limited by its very nature. This leads us to the fundamental question of the limitations of rational thought, and calls for a thorough examination of reason as a tool of inquiry into truth and knowledge. In other words, in our search for truth, we have first to establish how far we can proceed by the use of logic.

A logical system can never be constructed in the void and always has need of a foundation. There are, first of all, undefined concepts or primitive notions that are intuitively understood but remain undefined. There are, in addition, propositions or sentences that are to be taken as the axioms or postulates of the system. The undefined concepts together with the axioms may be said to serve as the premises of the system. Finally, there must be some rule(s) of inference, or rule(s) of logic, to put these premises into motion, so to speak.

In the example given earlier of the undefined concepts in geometry (point, line, and plane), attempts to describe the point as that which has no dimensions, that is, no width, no length, and no depth, are only descriptions to help us intuit the concept; if taken as definitions, they obviously beg the question as to what the definition of dimension is, implying merely a shift of terms that are to be taken as undefined—that is to say, a shift of undefined terms, but not the elimination of the need to accept some notions as undefined. Likewise, in Euclidean geometry, for example, nothing can be done in the absence of Euclid's five postulates, which are the axioms of the system. And, of course, unless there is a rule of inference, namely, mathematical deduction, no theorem can be proven.

Logic is a useful tool for the coordination of mental activity and for effective demonstration, but in itself it is not the source of truth because reason cannot function in the absence of premises and rules, the truth of which falls outside its scope.[9] Reasoning alone cannot arrive at truth, because man would not be able to recognize the truth arrived at as truth if he did not possess the essence of this truth *a priori*. Rational thought is thus inherently limited as an agent of the discovery of truth and, moreover, it points to this limitation itself by the very fact that, in order to be able to function, it needs to take the truth of its premises and rules for granted.[10]

Moreover, logic does not claim to give us a full account of truth by its own means. On the contrary, careful use of logic makes it clear that it is incapable of doing so. A reflection of this inherent limitation of rational thought is found on the level of formal logic:[11] every (non-trivial) formal system

9 In being logical or rational, we are innately seeking the reasons or "causes" of things. Logic, therefore, reflects within our mind a more universal reality, that of the universal causality which pervades all existence. We know that logic is valid because the law of causality is ingrained, as it were, deep in our spirit.

10 A reflection of this inherent limitation of reason can be found in mathematical logic in the form of Tarski's undefinability theorem, which holds that any consistent formal logical system, of sufficient complexity to render it of real interest, cannot define a truth predicate, that is, it cannot provide, by its own means, a criterion based on which a given proposition within the system will be decided true or false—truth in this context not being truth as such, but relativized to a given formal system. The undefinability theorem does not prevent truth in one system from being definable in a stronger system that would include primitive notions, axioms, and rules absent from the original system so that there are theorems provable in the stronger system that are not provable in the original system. However, this higher system would only be able to define a truth predicate for the original system. To define a truth predicate for the higher system, one would need a still higher system, and so on.

11 Formal logic bears the qualification "formal" because it does not deal with the contents of logical propositions, but rather with their forms and relationships. For example, the logical propositions "all swans are white," "all

that is constructed using a set of axioms will always include questions that cannot be answered by the means of that same system;[12] furthermore, such a formal system cannot prove, by its own means, that the system in question is consistent, that is to say, logic cannot prove its own logicality.

In this context, it is instructive to review an interesting episode in the history of mathematics, which, with all the checks and balances that its rigorous methodology provides, gives us a degree of confidence that no experimental science can possibly confer. The problem of securing a foundation for mathematics raised much interest among mathematicians at the beginning of the twentieth century. The aim was to organize all aspects of mathematics in such a way as to form a base of the most fundamental concepts, assumptions, and principles so that all other aspects would depend on this base. It was thought that all mathematics might be derivable from one such base. In 1922, Hilbert launched a foundational program, in order, as he said, "to settle the question of foundations once and for all." In 1931, Gödel proved that Hilbert's dream was not realizable with his proof of a pair of theorems, called the first and second incompleteness theorems, which, together with Tarski's undefinability theorem presented a few years later, set a strict limit on the power of logic, thus reflecting on the plane of mathematical logic the fact that our

men are mortal," and "all bears are brown" all have the same form but different contents and truth values.

12 The existence of a self-negating statement in every logical system of sufficient complexity is an indicator of the inherent limitation of the system in question, that is, it indicates that the system cannot fully judge itself from within itself and by its own means. This is an inherent limitation of formal logic. Gödel's non-provable sentence that "asserts its own non-provability," Turing's undecidable problem that "embeds its own undecidability," Russell's paradoxical set "of all sets that are not members of themselves," and so on and so forth, all reflect the same underlying idea. Rational thought shows its limitation when it is tasked to be the sole judge of itself, that is, by rational means.

rational faculty cannot be the sole judge of itself by its own means, that is, by reason.

That logic is limited in its scope and, as such, cannot serve as the sole criterion of truth was obvious to the ancients, who did not need the results of twentieth-century mathematical logic to know this. That being said, given that truth can always find a reflection in clear logical thinking, mathematical logic too can be a clear mirror of higher truths on its own limited plane. While the foundational works of Gödel, Tarski, and others do not offer the ultimate truth about the limitations of rational thought, they are nevertheless reflections of that truth on the plane of mathematical logic.

To repeat, man's rational faculty points to its own limitation. It thus indicates to man that there is "something"[13] deeper in him than reason. The very ability of reason to point to its own limitation, and to point to that "something" in man that is deeper than rational thought, shows that it takes its light from that same "something";[14] it is indicative of its being a reflection of that "something" on the plane of the mind. That deep-seated "something" in man, when it is reflected on the plane of the human mind, is refracted, as it were, into differentiated mental faculties, of which reason is only one.

The fact that man intuitively understands primitive concepts that cannot be defined rationally (the point, line, and plane, in geometry, for example) is indicative of a mental faculty besides reason, a faculty that may be termed intuition or

13 Meister Eckhart: "There is something in the soul which is uncreated and uncreatable . . . and this is the intellect" (*The Complete Mystical Works of Meister Eckhart*, translated and edited by Maurice O'C Walshe, revised with a foreword by Bernard McGinn [New York, NY: The Crossroad Publishing Company, 2009], p. 28).

14 According to St. Thomas Aquinas, the principles of logic are eternal, and as such reside in the Divine Intellect (*Summa Theologiae*, I, Q. 10, Art. 3, ad. 3).

insight.[15] Furthermore, the human mind is also endowed with imagination and memory. These faculties—reason, intuition, imagination, and memory—are the principal functions of the human mind. They all take their light from a deep-seated "something" in man, of which they are merely differentiated reflections on the plane of the mind.

Mental faculties are reflections of a deep-seated, limitless source—but on a limited plane, that of the human individual, precisely, who of necessity is finite. Consequently, the human mind is limited. But because it takes its light from that limitless source, it is able to point to its own limitation, and also to its limitless source—provided, of course, that it is not biased by arbitrary assumptions and prejudices.

As limited as the human mind is, it does not function mechanistically. On the contrary, man's capability for discernment surpasses all mechanistic or algorithmic systematizations, precisely because man's mental faculties take their light from a limitless source within him that cannot be confined to mechanistic systematizations. Intuition or insight, for example, cannot be mechanistically systematized.

The fact that intuition or insight is a necessary element in correct discernment is reflected in the problem of undecidability on the plane of logic:[16] many important problems in mathematics and logic are shown to be undecidable, that is, it has been proven that no effective algorithmic method for deriving the correct answer to them can exist.[17] To be clear:

15 It is this same mental faculty that enables man to conceive of the paradoxical statements—Gödel's non-provable sentence, Turing's undecidable problem, and Russell's paradox, for instance—that are keys to the conception and proof of the foundational theorems of mathematics, of which mention was made above.

16 In this regard, one is referred to the works of renowned mathematicians Alan M. Turing and Alonzo Church.

17 Other than elementary propositional logic and similar very restricted systems, formal logic in general is undecidable.

it is not a question of finding algorithmic methods, but the conclusion that such methods in principle cannot possibly exist. In other words, the existence of undecidable problems is not a shortcoming of particular formal systems, but rather a property inherent in nearly all formal systems. The importance of this observation is far-reaching. For example, it puts to rest all claims about the reconstructibility of the human mind by means of artificial intelligence,[18] and all the pseudo-scientific speculations deriving from it that are sometimes indiscriminately accepted as scientific by the common man of our time.

## How Can We Know that a Doctrine is True?

The notions of science and truth are interrelated: science is synonymous with knowledge; and the latter, of course, is the knowledge of truth; otherwise, it would not be knowledge, but ignorance. As noted, logic cannot be the sole criterion of truth. How can we know then that anything is true? What should be our criteria for distinguishing truth from falsehood?

Logic demands that the axioms of a system should be consistent; otherwise, a statement and its negation will both be derivable from those axioms, thus rendering the system useless. Moreover, a system of thought that sets out to treat a subject should be able to account satisfactorily for the things that fall within its scope. That is to say, its set of axioms should be adequate to make for a scope wide enough that will include the things it wishes to describe. Otherwise, it will not fulfill the goal it has set for itself.

For example, absolute geometry, namely, a system of geometry that has the first four of Euclid's five postulates as its

18 As noted, for example, in J.R. Lucas, "Minds, Machines and Gödel," *Philosophy* (1961), Vol. 36, No. 137, pp. 112-127, and in Roger Penrose, *The Emperor's New Mind* (Oxford, UK: Oxford University Press, [1989] 2016).

axioms, is consistent but is not sufficiently strong, that is, it is not adequate to describe all geometry as we generally know it; it remains silent about many theorems that Euclidean geometry proves, namely, all those theorems that require the parallel postulate as an axiom. For instance, it neither proves nor disproves the theorem, provable in Euclidean geometry, which states that the sum of the interior angles of a triangle is two right angles. Newtonian mechanics is another example of a system with a limited scope: it satisfactorily describes the motion of objects normally within our reach; however, it fails in the face of certain more sophisticated observations, which can be accounted for by more adequate systems, such as quantum mechanics.

So long as a consistent system of thought does not present claims that fall outside its scope, it remains consistent, albeit potentially inadequate. If, on the other hand, it presents a claim that is not supported by its axioms—that is to say, it presents a claim that does not fall within its scope—it will become inconsistent, precisely because the negation of the same claim can equally be made without for all that contradicting the axioms of the system. Therefore, a consistent yet inadequate system will lose its consistency as soon as it breaches the boundaries of its scope by making a claim that is not supported by its axioms.[19] For example, absolute geometry is silent about the triangle angle sum theorem while remaining consistent within its scope; it would become inconsistent if it pronounced a judgment about this theorem. Rationalism, to take another example, refutes itself (as we

---

19 We note that a consistent yet inadequate system that makes a claim that is not supported by its axioms is no longer the original consistent system: it is a new system that has the axioms of the original consistent system plus one more implicit axiom—the axiom, precisely, that holds that the said claim is supported by the axioms of the original system; the presence of this unacknowledged axiom renders the new system inconsistent. For example, when empirical science, consistent and effective within its own scope, expresses opinions about immaterial states, it does so inconsistently.

have demonstrated in the Introduction) precisely because it presents a claim that falls outside the scope of reason: the claim, namely, that reason is the sole criterion of truth—a claim that reason itself cannot prove. The overarching principle of rationalism—that reason is the criterion of truth—is not rationally provable, and, as such, will have to be taken for granted outside any reasoning and in direct contradiction to this very principle.[20] As we saw in the preceding section, no sufficiently elaborate logical system can prove itself true from within itself, that is, by logical argument. Consequently, when assessing a theory, what we *are* able to do by logical means is to try to prove its premises inconsistent by finding statements formed from those premises, and in accordance with the rules of logic, which would contradict the said premises. If the theory is thereby proven inconsistent, it will be of no value. What we *are not* able to do is to prove a theory true from within itself; for instance, we cannot prove, by our reason, that a theory about our mind is true.

If logic cannot provide us with the proof of the truth of a theory, then, in our search for truth, what should be the basis of our acceptance of a given theory? To remain logically sound, the theory should not contradict itself. That is as far as logic can go; in other words, that is the outer limit of rational thought. One cannot logically expect all explanations that a theory provides to be logically provable—precisely because that is beyond the claims of logic; one can only expect them not to be proven false when rigorously tested. Moreover, as previously noted, the theory should be adequate in accounting for what falls within its scope. Consequently, unless a theory is proven either inconsistent in its premises or inadequate in its scope by way of evidence, we have no reason to deem it invalid or inadequate.

20 Rationalism, therefore, must be distinguished from rationality. Rationalism seeks to confine man's faculties of knowledge within the bounds of reason; it is the illogical exclusivism of reason, not the mere use of it.

Popper's principle of falsification and the criterion of consistency resemble one another in that they are both based on the observation that, contrary to the positivist claim, truth cannot always be proven true and that a theory is acceptable if it cannot be proven false; that is to say, an acceptable theory should remain consistent in the face of all careful examination. In both approaches, therefore, one is conscious of the inherent limitations of reason—namely, that logic cannot prove every true statement true, whereas it can prove an inconsistent theory false, precisely by showing its inconsistency. In another respect, however, the two approaches are different: falsificationism demands that a scientific theory must allow for an experiment whereby, in principle, it could be proven false; we argue, on the contrary, that while proposing experimental methods for the empirical sciences is plausible, limiting science as such to the domain of sensory experience is arbitrary. Unlike Popper's principle of falsification, the criterion of consistency does not require an experimental framework. In short, Popper's principle of falsification will coincide with the criterion of logical consistency—a well-known criterion to the ancients—when it is stripped of its empiricist assumption, which in any case is a logically-unsupported extrinsic "dogma" as far as science as such is concerned. Popper, of course, was careful not to lay claim to total truth or to science as such: he consciously limited the scope of his investigative principle to the empirical sciences. However, the tacit empiricist premise of modern science is often ignored, unconsciously or consciously, and as a result, "science as such" and "empirical science" are often used interchangeably. The "empirical assumption" renders science inadequate and the "empiricist assumption" renders it inconsistent: "empirical" science is useful only within a limited scope, that of sensory experience, and "empiricism" is logically inconsistent since the principle of empiricism is not logically or empirically provable and is thus unscientific by its own measure. Either way, one who is in search of truth should look for something other than what modern empirical, or empiricist, science offers: so long as a theory is not proven inconsistent or inadequate by careful examination, to remain logical, a reasonable man must not consider it unacceptable. As we shall see in the next two chapters, traditional doctrines of epistemology and metaphysics have never been proven inconsistent or inadequate, and in rejecting them in the absence of any evidence to the contrary, the rationalist has only proven himself irrational.

In summary, this will be our guideline in judging the truth of any doctrine: to examine its acceptability, we shall study its consistency and adequacy. In fact, we have already used these two criteria more than once in our preceding arguments. In conformity with the criterion of consistency, we refuted rationalism, subjectivism, relativism, and agnosticism by showing their intrinsic inconsistency; likewise, we showed that every attempt to define science will be self-refuting, that is, it will result in contradiction; furthermore, we showed that inductivism, empiricism, and positivism all carry the seeds of their refutation within themselves and are thus inconsistent. And in conformity with the criterion of adequacy, we noted that while the use of induction, empirical investigation, or logical deduction—as tools of limited scope in their application—does not lead to inconsistency, these tools are inadequate as far as science as such, or total truth, is concerned. Furthermore, as we shall see later in this book, the modern perspective on science falls radically short of satisfying the criterion of adequacy subjectively as well as objectively—subjectively, by its unwarranted limitation of the faculties of the knowing subject to the sensory faculties and reason, and objectively, by its unjustifiable restriction of the object of knowledge to matter.

### The Subject and Object of Science

Science, inasmuch as it is knowledge, necessarily involves a subject that knows and an object that is known. Inseparable from the notion of science, therefore, are the nature of the knowing subject and that of the known object. Accordingly, an important question in the discussion of the nature of science is that of the faculties of knowledge in the human subject: what are the faculties of man that enable him to perceive the objects of knowledge? Another equally important question is that of the nature of the objects of knowledge: what can man, in principle, know? Answers to these two questions form the foundation of an integral philosophy of science.

Our human subjectivity perceives the objects of the material world by means of its sensory faculties, and it reasons with its rational faculty. The inherent limitations of man's rational faculty discussed earlier show that the human subject possesses a faculty of knowledge that goes beyond logic; this deep-seated faculty is what allows man to recognize and acknowledge the truth of a logical argument, for example. The question of the faculties of knowledge in man is the central theme of epistemology, which is traditionally considered to be a fundamental science, and, as such, is worthy of careful study; to that we shall turn in some detail in the next chapter. Each of the degrees of man's faculties of knowledge has its own object of knowledge that belongs to a corresponding state of being. Consequently, there are multiple degrees of being, each corresponding to the state of the object of knowledge of a given degree of man's faculty of knowledge. Metaphysics, the study of the multiple states of being, or the multiple degrees of reality, constitutes another traditional science, which, like epistemology, is fundamental and thus worthy of careful study; of that we shall present a detailed account in a later chapter of this book.[21]

21 In chapter 3, "The Degrees and Modes of Reality."

## CHAPTER 2

# MAN'S FACULTIES OF KNOWLEDGE

Epistemology, the science of the nature of man's faculties of knowledge, is fundamental to the understanding of the way man perceives all science. Epistemology relates to the knowing subject while the doctrine of reality—metaphysics—relates to the object of knowledge. The main question that epistemology aims to answer is this: what are man's faculties that enable him to know? Any integral philosophy will of necessity have an epistemology. Platonic *anamnesis* is perhaps the most explicit epistemology in the ancient West, and is in full agreement with the epistemological doctrines of all the great religions, which may vary, however, in their form of expression or in their degree of explicitness.

As we saw in the preceding chapter, man's faculties of knowledge include the sensory faculties, by which he perceives the objects of the material world, and reason, which is itself a reflection, on the plane of the human mind, of a more deep-seated faculty—the intellect, that "uncreated and uncreatable" element in man which allows him to recognize and acknowledge the truth of all that is true. Thus, there is, in the hierarchy of man's faculties of knowledge, in addition to the senses and reason, the element intellect whose operation—intellection or intellectual intuition—is referred to as "recollection" or *anamnesis* in Plato's theory of knowledge.[1]

---

1 "It is said that the senses are powerful. But beyond the senses is the mind [*manas*], beyond the mind is the intellect [*buddhi*], and beyond and greater than the intellect is He [the supreme Self]" (*Bhagavad Gītā*, III:42, Shri Purohit Swami translation).

The existence of the sensory and rational faculties is hardly disputed in modern philosophy. On the other hand, the intellect—that divine spark of which all other cognitive faculties are merely reflections—is often ignored, or denied, in modern thought, and, as such, it deserves particular consideration. It is to this that the rest of this chapter is devoted.

### Platonic *Anamnesis*

Is it possible for man to know anything? If yes, how can he know it? Plato answers these questions by proposing that "seeking and learning are in fact nothing but recollection."

We find the first treatment of this subject in Plato's *Meno*, where Socrates addresses the Sophists' claim that learning is impossible and inquiry futile, since man can neither learn what he already knows nor can he learn what he does not know: it is futile to inquire about what he already knows, since he already knows it, and if he does not know what he is looking for, then surely he will not be able to find it. But Socrates argues that there is an intermediary state between pure knowledge and pure ignorance, and that it is this: man knows the object of inquiry but has forgotten it; he will remember it when he sees it.

But on what basis is this *anamnesis* possible? If a man learned something in his lifetime and forgot it later, the question will then be: how did he learn it in the first place? This can only be resolved if we assume that he knew it before he came to this world, that he knew it from a former life in another world. Here terms such as "a former life" and "another world" prior to this world are not to be taken literally,[2] but are employed merely to indicate, metaphorically, that principial realities are contained in man's spirit as potentialities or virtualities so that when he sees their rays in this world, he remembers them.

---

2 Otherwise, the question will arise as to how he learned it in his former life.

Socrates demonstrates his theory by posing questions to a boy in Meno's entourage who knows nothing of geometry. Socrates asks the boy if he knows how to draw a square double the size of a given square. After a few unsuccessful attempts by the boy, Socrates draws a second square, its side on, and equal in length to, the diagonal of the initial square. With some hint from Socrates, the boy realizes that in this way a square with an area twice as large as that of the initial one is constructed. Socrates then notes that the boy is able to discover this truth of geometry because he knew it *a priori*. It follows that this knowledge was innate to him and that Socrates' role has simply been to help him remember it. The teacher's role, then, is merely to awaken in his students the knowledge they already possess deep in their very being.

Much criticism has been leveled at Plato on this subject, one being that Socrates induces his own ideas into the boy and then claims that the boy discovers them. The answer is simple: granted, it is Socrates who instills the idea in the boy, but who instilled the idea in the person who discovered this theorem in the first place? Unless one accepts the postulate that knowledge is innate to man, all reasoning is flawed.

Plato returns to this subject again in his *Phaedo*. As proof of the immortality of the soul, it is first proposed that our birth is a sleep and a forgetting, and that to learn is in fact to remember the knowledge that must have been received in another life, which implies that our life did not begin with our earthly birth, and likewise does not end with our bodily death. Socrates, here, presents an even more profound reasoning for the immortality of the soul. The soul is immortal because it can perceive eternal ideas such as truth, beauty, and goodness. Man can know God because he has, deep within himself, something eternal that does not die.

To demonstrate his thesis, Socrates gives the example of the notion of equality-in-itself, which is independent of any particular case of equality perceived by the senses such as equal sticks or equal stones. There are no instances of an absolutely perfect equality in the sensible world, and yet we have had this notion all our lives. Socrates concludes that we could not have come to learn of equality-in-itself through our senses, but that we gained its knowledge before our birth. These fundamental ideas never die and can never be learned in the void. They are contained in our spirit, and are awakened in us when we see their manifestations in the world.

Platonic epistemology has been formulated in many different ways, which are all essentially the same: the existence of fundamental ideas or principial realities in the human subject, which are not obtained, and not obtainable, by experience, is the proof of their objective reality; we can learn them, that is, we can remember them, because we possess them deep in our heart.

The ontological proof of God is based on the same idea, namely, that the very fact that we can conceive of an absolutely perfect being is the proof of its objective reality.[3] This too has been the subject of much criticism: it is objected, for example, that one can think of a winged horse, yet such a creature does not exist. But this is to forget the difference between funda-

---

3 The ontological argument has been formulated in many different ways. St. Anselm's formulation, an argument by *reductio ad absurdum*, may be summarized as follows: Any man can conceive of a being "that than which nothing more perfect can be conceived of." If that being existed only in the mind and did not have objective reality, he can then conceive of another being that has all the qualities of that being plus objective reality. This latter being, given that it has one more positive quality than the former being, is even more perfect than it, that is to say, it is more perfect than "that than which nothing more perfect can be conceived of," which would then lead to logical absurdity. Therefore, "that than which nothing more perfect can be conceived of" cannot exist only in the mind and has to be objectively real. We do not pretend that this logical argument is without premises, but we note that *any* rational argument of necessity has premises. One who asks for a rational argument, to be logical, has to accept the assumption of some premises. The ontological argument is not, however, without merit: it is a logical demonstration that can have an awakening power for some. For certain others, beauty, for instance, can serve the same function.

mental ideas and the objects of sensory experience. It is clear where we get the idea of a winged horse from: a horse is an object of our experience, as is a wing. Consequently, a winged horse is not a fundamental idea that we would carry within us independently of sensory experience. As such, the Platonic argument is not applicable to it, and the objection is not valid.

Convincing as it is, Platonic *anamnesis* is not thus far *proven* true, and remains a theory—that is, as far as logic is concerned. But can we prove it true by means of reason? We cannot. Logic cannot prove every truth, as we have shown extensively in the preceding chapter, and consequently it can have no claim to total truth. What then should be done? If we are not able to prove Platonic *anamnesis* by means of logic, what is the basis for our acceptance of it? As discussed earlier, in order for a doctrine to be acceptable, it should be logically consistent and it should be able to account adequately for the things that fall within its scope. As such, we have to accept the Platonic epistemology unless someone can successfully invalidate it either by proving it inconsistent or by showing it inadequate.

### Has Platonic *Anamnesis* Withstood Attempts to Prove it False?

Much has been done over the past few centuries to attempt to invalidate the Platonic theory of knowledge, or to offer alternatives to it. Our goal here is not to provide a comprehensive account of all such attempts—their worth would not warrant such prolixity—but to present some of the most well-known criticisms leveled at Platonic *anamnesis*, and to briefly assess their validity.

Descartes, the father of modern philosophy, who defines knowledge in terms of doubt, is logical enough not to deny the presence of innate notions in the soul, but in practice he restricts them to the domain of reason. However, reason itself does not authorize such a restriction: rationalism is intrinsically inconsistent, as we have discussed in the preceding chapters.

As an advocate of empirical criteria for knowledge, Hume holds that there are no innate ideas, and that all knowledge comes from experience. He is known for having applied this idea to causality. Instead of taking the notion of causality for granted, as Plato would, Hume attributes it to experience. And this he would have us believe: Of two events, we say one causes the other when the two always occur together. Whenever we find one, we find the other also, and we are thus psychologically conditioned to believe that the latter will follow the former. The sun rises and the rocks heat up, for instance. We psychologically associate the idea of the rocks becoming warm with the rising of the sun. The relationship between the two is not causal, it is merely an association of ideas in our minds. The fallacy of Hume's argument is obvious: in explaining what produced the notion of causality in our mind, he is effectively trying to ascertain the cause of causality. In other words, he must have taken causality for granted to have been prompted to explain its cause in our mind.[4]

For Kant, man of necessity experiences things in determinate localities in space and time, and he cannot but perceive things in certain ways—through causality, for instance. Man, he would have us believe, is imprisoned in his subjectivity: he has no way to know if these things have objective reality or

---

4 We find the notion of causality in us because it is a universal reality that is necessarily present in our deepest being. When reflected on the plane of the mind, this innate idea manifests itself as our need for the causes or "reasons" of things, thus empowering us with our rationality, of which logic is a clear expression. Causality, therefore, is a more fundamental reality than logic: the latter is merely the reflection of the former on the plane of the human mind. To accept an effect without a cause, then, is as absurd as to accept an inconsistent thesis. He who does not accept causality should best remain silent, for he would find himself in refutation of himself as soon as he expresses an opinion.

not. Now, Kant does not present any reason why man's situation is such as he describes it. This in itself, however, is not a refutation of his thesis. We have first to assume it true and see if it holds, that is, if it remains consistent. If man is subjective and has no way of knowing anything objectively, then how did Kant come to know that man lives in his subjectivity and is confined to it? In other words, how can a man imprisoned in his own subjectivity proclaim an objective truth about everyone, including himself?[5]

In his *History of Western Philosophy*, Russell, while implicitly admitting that there is no logical contradiction in Plato's theory of knowledge, casts doubt on its universal applicability by trying to show that if Plato's theory were indeed true, one should be able to receive empirical knowledge, namely, things that are solely at the disposal of the sensory faculties—historical or geographical data, for example[6]—from within oneself without the aid of one's sensory faculties; in that case, divination, which Russell regards as pure superstition, should also be possible.

To respond to Russell, one will first have to note that Plato does not claim that every man remembers what he knew. *Anamnesis* bases the possibility of knowing an object on man's having its knowledge potentially. It does not say that this knowledge is actual in all men; it says that *every* man possesses it deep within his being while only *some* actually

5 Kant is not alone in declaring subjectivism as a universal and objective truth. In modern psychology, for instance, the knowing subject is posited as purely psychic, and nothing more: "The psyche is the object of psychology, and—fatally enough—its subject at the same time and there is no getting away from this fact" (Carl G. Jung, *Psychology and Religion* [New Haven, CT: Yale University Press, 1938], p. 62). According to this opinion, every man is "fatally" imprisoned in his psyche except for Jung, who has "gotten away from this fact" to inform us of it.

6 According to Russell, the boy in Meno's entourage "could not have been led to 'remember' when the Pyramids were built, or whether the siege of Troy really occurred, unless he had happened to be present at these events."

remember it. To prove the Platonic thesis false in Russell's way, it will be sufficient to find one man who cannot possibly learn anything—nothing at all. Anyone who speaks or has any way of communicating something to us has learned how to communicate, so he will not qualify as a rebuttal witness. A blind, deaf, and mute person, apparently a good rebuttal witness in this case, would have no way of telling us that he can learn nothing. Yet Helen Keller is an astonishing example to the contrary.

But to return to Russell, he thinks it impossible for a man to have access to empirical data without the use of the sensory faculties, or to foretell future events. Now, logically, to be able to prove his point, he would have to show that no man can accomplish such feats—an obviously impossible task for Russell to do.[7] On the contrary, the following case provides an astonishing example of the possibility, precisely, of what Russell thought obviously impossible, that is, of having the knowledge of historical and geographical facts without the use of the sensory faculties.

At the beginning of the nineteenth century, Anne Catherine Emmerich, a bedridden nun in Germany, had a series of visions in which she saw the life of Christ, especially his last days, as well as details of the life of the Virgin Mary. Clemens Brentano transcribed her visions, which were made into two books published after her death.[8] One of Emmerich's accounts contained descriptions of the house that St. John the Apostle built near Ephesus for the Virgin, and where she lived to the

---

7 We note that the burden of proof is on Russell. Even in the absence of the examples given here, his refutation is logically invalid. These examples, though not logically necessary, are meant to provide "empirical" evidences attesting to the very opposite of his opinion.

8 *The Dolorous Passion of Our Lord Jesus Christ According to the Meditations of Anne Catherine Emmerich*, first published in Germany in 1833, and *The Life of the Blessed Virgin Mary from the Visions of Anne Catherine Emmerich*, first published in Germany in 1852.

end of her life. Emmerich provided much detail about the location of the house and the topography of the surrounding area. She also gave details of the house itself. Prior to Emmerich's visions, people generally thought that the Virgin lived in or near Jerusalem to the end of her life. However, towards the end of the nineteenth century, some priests, relying on Emmerich's visions, searched and discovered a small stone building on a mountain overlooking the Aegean Sea and the ruins of ancient Ephesus in Turkey, which perfectly matched her descriptions, and which had been venerated for a long time by some of the locals, who were descendants of the early Christians of Ephesus. This house is now widely venerated as the house of the Virgin.

As for the possibility of foretelling future events, which Russell considers so unlikely,[9] there is no logical contradiction in a model that accepts an omniscient God who knows of events and destinies in advance, and who could inform some individuals through angelic inspiration. Pharaoh's dreams and Joseph's interpretation of them, related in the Bible and the Koran, is a case in point. In fact, examples of the kind, on various levels, abound not only in religious texts but also in numerous non-religious works.

We thus see that none of the philosophers we have referred to here have refuted Plato. Needless to say, refusing to accept a doctrine is not the same as refuting it. For the Platonic recollection of the fundamental truths to become actualized in a man, a purification of the heart is necessary—in order that earthly shadows may awaken in him their principial realities. The aim of all the religions is to help man remember what he knew—to remove the rust that covers his heart, symbolically speaking—so as to enable him to reestablish his contact with the truths contained in his inner being. Where

---

9 We note that this does not relate directly to Platonic *anamnesis*; our point is that Russell's objection does not pose any logical challenge to the possibility in question.

the layer of the rust on the heart is thick, however, logic is of little avail, even to a logician such as Russell.[10]

Platonic *anamnesis*, we see, solidly withstands these attempts to prove it false. On the contrary, every careful and unbiased examination points to its truth. Renowned linguist Noam Chomsky's careful studies are perhaps the most recent investigative testimony to the truth of Platonic *anamnesis*. In Chomsky's view, man's language faculty contains innate knowledge of various linguistic rules, constraints, and principles, which, through interactions during the life of a child, gives rise to the knowledge of specific languages. What makes language acquisition possible is the fact that much of our linguistic knowledge is unlearned, that is, it is innate to us.

---

10 "To whomever God showed not the way / to him naught was opened by the use of logic," sings Mahmoud Shabistari, an eminent fourteenth-century Sufi sage (*The Garden of Mystery*, 87).

CHAPTER 3

# THE DEGREES AND MODES
# OF REALITY

Closely related to any theory of knowledge, that of the faculties of the knowing subject, is a theory of being or reality, that
of the nature of the objects of knowledge. Modern science, by
and large, ignores those phenomena which are not within the
reach of man's sensory faculties. There is no logical justifica-
tion, however, for this restriction of the scope of the objects of
our knowledge. On the contrary, if our faculties of knowledge
are not limited to our senses, as we noted is the case in our
discussion of epistemology, our doctrine of being should likewise be sufficiently broad in its scope to account for all that, in
principle, is knowable to man.

What are the criteria for an acceptable theory of being?
We noted that any consistent system of thought that sets out
to treat a subject should be able to account satisfactorily for
the things within its scope. Accordingly, if a doctrine of being
is proposed, and its premises are not proven inconsistent, it
should be able to explain all phenomena of being, that is, it
should be able to account for all beings in such a way that
cannot be refuted by evidence.

## The Degrees of Reality

The notions of reality and being are interrelated. That is *real*
which *is*, and that which *is* is thereby *real*. Therefore, to speak
of reality is to speak of being, and vice versa.

The perspective of the great traditions on the doctrine of
being differs widely from that of modern man in that the tra-

ditional outlook always accounts for multiple states of being, each corresponding to the objects of a degree of man's faculty of knowledge. In contrast to modern science, for which the material world is all there is, the traditional doctrines of reality account for immaterial beings, of which the religions so often speak—angels and spirits of all kinds, for example.

There is, however, no unique classification of the various degrees of being, and depending on the scale of gradation desired, different accounts may be given. Some, considering the essential identity of Creation with the Principle, may describe total reality as being one indivisible Unity. Some may divide it up into two degrees: the Divine Order, and that of all that is created, namely, Creation. At the other end of the spectrum, some, in order to account for a more nuanced classification, have proposed forty states, for example. An intermediary between the two ends of the spectrum, one sufficiently nuanced to provide fundamental differences while maintaining an appropriate level of conciseness, is the doctrine of the five Divine Presences,[1] which is the basis of the presentation here of the doctrine of the multiple degrees of being.

Starting from the material world that is in principle knowable to our senses, we perceive the first state of being, that is, the material state that surrounds us, encompassing the entire system of galaxies, however inaccessible to us they may be in fact. To some, this state represents all there is. We shall see, however, that this supposition radically falls short of adequacy.

Our thoughts, for example, have an immaterial reality. The same is true of a melody that is conceived in its form but is not yet physically played out. Thoughts, melodies, and other realities of this kind, all have forms: a statement in a human language, for example, has a structure, as does a melody. Consequently, they all belong to a state of being that is

---

1 Adopted, in its essence, from Frithjof Schuon, "The Five Divine Presences," *Form and Substance in the Religions* (Bloomington, IN: World Wisdom, 2002), pp. 51-68.

formal. The form of a sentence or a melody, however, is of a different kind from that of a tree, for instance. One may say that the melody's form is more subtle, or less gross. Accordingly, we shall refer to the material (and sensorial) state as the gross formal state, and to all that is formal, yet not gross, as the subtle formal state.

The entirety of the preceding two states—the formal state—is that of the individual: one tree is different from another, for example, as is one individual's thought from another's. One individual is distinct from another, and, as such, its form excludes those of others. To say form, therefore, is to say exclusion of other forms. Formal logic falls within the scope of the subtle formal state, and, as such, formal systems exclude one another. But the truth of logical reasoning, for example, is universal. What makes us acknowledge the validity of a logical argument, the proof of a theorem, for example, is our access to universal truths underlying those arguments. The fact that a logically valid reasoning by one individual is convincing to another shows that both individuals possess truths that are thus independent of the individuals, that is, they are universal. These truths, when refracted in the human mind, crystalize into formal formulations, but are in themselves formless. Universal realities do not have forms, yet formal realities have their roots in them; the latter are, as it were, the coagulations or crystallizations of the former. One, therefore, refers to universal realities as formless, and better still as supra-formal since they are the source of their formal crystallizations. Thus, in addition to the two formal states, gross and subtle, we distinguish within Creation, a supra-formal state encompassing the principles of all that is formal.

The relationships among the three states within Creation, namely, the three cosmic states since they relate to the states of being in the Cosmos, may be better understood by use of symbols taken from the material world. The substance of the gross state is likened to earth, that of the subtle state to fire, and that of the supra-formal state to light. Needless to say, all these three substances, namely, earth, fire, and light, belong to the material world, that is, to the gross formal state. The vertical relationships are thus reflected on a horizontal plane, only to provide analogies for a better understanding of these notions. One finds in a lit candle all three: the body of the candle would represent the gross state, the flame would represent the subtle state, and the light emanating from the flame would represent the supra-formal state. The candle is made of a gross substance, that is, relative to those of the flame and light, and has a form. The flame too has a form but it is made of a more subtle substance. The light, however, is formless. The light from one candle does not exclude the light from another. Wax, plasma, and light all belong to the material world, as we know, but if their horizontal relationships are transposed to a graded vertical dimension, they can be indicative of the realities of the gross, subtle, and formless states, symbolically speaking.

Man, this small mirror of the whole creation, finds within himself all three degrees of being in the Cosmos, that is, the gross, subtle, and supra-formal states. As such, in comparison to the Macrocosm, which is the world taken in its entirety and encompassing the aforementioned three degrees of existence, man is called the microcosm. His body (*corpus* in Latin, σῶμα in Greek) belongs to the gross state, his soul (*anima* in Latin, ψυχή in Greek) belongs to the subtle (or animic) state, and his spirit (*spiritus* in Latin, πνεῦμα in Greek), or equivalently his intellect (*intellectus* in Latin, νοῦς in Greek), belongs to the supra-formal state. The seat of the intellect is symbolically the heart, in contradistinction to the seat of the mind, which is symbolically the brain.

Supra-formal realities reflect within Creation principial realities that are uncreated and imperishable. These are possibilities. Birds fly because flight is possible or, otherwise said, because the possibility of flight *is*. Even if the world and all the flying beings it contains perish, the possibility of flight will remain unaffected in its principle. Possibilities, therefore, have a reality superior to that of their reflections in Creation. Unless something is possible, it cannot come into existence.[2] In other words, prior to existing, the possibility of a thing or a being must *be*. Existentiation, then, is only a "downward," or "outward," projection of possibilities. A creature is the manifestation in the world of a reality of a higher order. Creation, Existence, Manifestation, the Cosmos, the Universe, and the World[3] are, therefore, more or less synonymous, and represent a projection *ab extra* of higher realities, that is, of possibilities or ideas. In this sense, man's spirit or intellect—the "uncreated and uncreatable" element in his soul, in the words of Meister Eckhart—is the bridge that connects him to principial realities. Man is capable of knowledge, Platonic *anamnesis* has it, because universal truths are inscribed in his spirit. The theory of knowledge thus rejoins its corresponding theory of being.

It is somewhat difficult to imagine realities that are independent of space and time, the containers of the existential state, at least given the way we perceive these containers. An example, therefore, may be instructive here. The Pythagorean theorem, to take an example from mathematics, holds true wherever and whenever there is a right triangle. Even before the theorem was discovered by a human subject, it had held true. It "manifests" itself in the world, as also in the mind of the human subject that conceives of the triangle, where and when a right triangle exists; as such, it manifests itself in space and time. If the whole world were to collapse into naught, the theorem would no longer be manifested, but in its principle it would remain unaffected. It will come into

2 Each world, of course, is a manifestation of the possibilities that are in harmony with other possibilities in the same world, that is, each world is a manifestation of compossibles.

3 The Arabic word for the "world" is a *nomen instrumenti* of the root "to know," and means "that which makes known," that is, that by which God makes Himself known; thus, in essence, it means Manifestation.

existence again when and where a right triangle comes into existence. But what happens to the principle of this theorem in the meantime? Possibilities come into existence, or manifest themselves in Creation, but their fundamental realities are of a higher order—an order that may be referred to as the Principle or the Divine Order, God. Creation, therefore, points to its metaphysical cause, the Principle, which contains all principial possibilities—of which the manifested things in all degrees of Existence are only reflections. Things of this world *exist* because their principial realities *are*—and are on a higher plane than that of Existence. The world *exists* because the Principle *is*. Strictly speaking, one cannot say that the Principle exists, because it is beyond Existence;[4] one must say, instead, that the Principle *is*. The degree of the Principle, the Divine Order, may thus be termed the degree of Being.

Principial realities are all contained in the Principle, that is, in the Divine Order, regardless of whether they reflect themselves in Existence or not. Everything that exists is a reflection of a possibility; otherwise, it could not come into existence, precisely. But the converse is not necessarily true; there are possibilities that cannot be manifested: God the Creator is a possibility that remains in the Divine Order. God's image in Creation, in the supra-formal state, that is, the Spirit of God, or the Divine Intellect, is the summit of Manifestation, but the original possibility remains *in divinis*. If God as such manifested Himself in the world, the world would burn to ashes.[5]

---

4 Existence, from *ex* (out) + *stare* (to stand), is a more outward state of being than Being.

5 "And he [God] said, Thou canst not see my face: for there shall no man see me, and live" (Exod. 33:20). Some may object that Christ—"God"—did enter into this world, which did not burn to ashes. If it is true that "the Word was God . . . and the Word was made flesh, and dwelt among us" (John 1:1-14), it is equally true that Christ said, "Why callest thou me good? there is none good but one, that is, God" (Matt. 19:17), which clearly establishes a distinction between the Principle *in divinis* and a direct reflection of it in Manifestation.

God the Creator is, so to say, the first self-determination of the unconditioned Essence, much as, metaphorically speaking, speech is a determination of silence, from which it springs forth and to which it returns. The Absolute—the Essence—admits of no qualification whatsoever; otherwise, it would not be absolute.[6] Any description necessarily implies a determination, hence a limitation. That is why the Supreme Principle, the Essence in itself, is often described in negative terms: the *Unconditioned Essence*, *Non-Being*, the *Impersonal God*, the *Void*,[7] whence the origin of apophatic or negative theology. This negation of attributes is not a privation, obviously: it is plenitude beyond description. To emphasize this point regarding the Essence, the term Beyond-Being may be used instead of Non-Being. Names of the Essence, such as the Absolute, Non-Being, Beyond-Being, the Impersonal God, the Supreme Principle, the Sovereign Good, and the Most Holy are not determinative attributes. Divine Attributes, on the other hand, such as the Merciful and the Forgiver [of sins] refer to different self-determinations of the Essence, hence to the Creator, Being, or the Personal God, since they refer to an extrinsic quality that has the creation in view: for unless there are creatures, there will be none, for example, to exert mercy upon or to forgive.[8] Thus, within the Divine Order, two degrees are distinguishable: that of Being, or God the Creator, and that of Beyond-Being, or the Essence.[9] The principial possibilities that prefigure created realities are dif-

6 The Essence cannot be man's interlocutor although man can become identified with it by extinction.

7 The *Shūnya*, or *Nirvāna* (extinction), of the Buddhists.

8 For the Divine Qualities to remain actual, there is always a "need" for creation. To say Principle is to say Manifestation just as to say sun is to say radiation.

9 The three degrees of Beyond-Being, Being, and Existence are very well expressed in German by the terms *Ursein*, *Sein*, and *Dasein*. It is the latter, let it be noted in passing, reduced to its most exterior shell, that of the lowest level of experience, which is the ultimate reality for the existentialists.

ferentiated on the level of Being; they are all contained in the Essence in an undifferentiated way.

Viewed from man's perspective, total Reality may be perceived as consisting of the Divine Order and Manifestation, the two then being subdivided into two and three degrees, respectively. The two degrees distinguishable in the Divine Order and the three states of being that refer to Manifestation, considered altogether, constitute the five Divine Presences. When the two formal states, the gross and subtle, are taken together, they form the "natural" domain or the domain of "nature."[10] Viewed this way, total Reality may be thought of as consisting of four degrees, the lowest of which will then be the formal state or the natural world.

Cosmology is the science of the Cosmos, and, as such, accounts for the three states of being that constitute the created world—in relation, of course, to the metacosmic Principle. Ontology is the science of being and includes, in addition to Manifestation, the ontological Principle, Being. Finally, metaphysics, the science of reality, includes, in addition to Manifestation and Being, Beyond-Being or the Essence; it thus encompasses in its scope all five degrees of reality.

Our terrestrial world is the material state that surrounds us; it encompasses all that is in principle knowable to our senses, on earth or beyond. The gross state, however, is not solely the terrestrial world we know, for the subtle state envelops myriads of other crystallizations comparable to our terrestrial world, but unconnected with it, and completely inaccessible to our sensory faculties. Similarly, viewed from the supra-formal state, there are many worlds of a subtle nature; and viewed from the Principle, the worlds of light which are the Paradises extend in an unimaginable profusion, like the drops of a fountain illumined by a ray of sunlight.[11]

---

10 This application of the word, we note, is in stark contrast to the materialistic view that limits the natural domain to that of matter.

11 See Frithjof Schuon, *Form and Substance in the Religions*, p. 67.

Man's reality, we have said, contains objects of all the three states within Creation: his body is an object of a gross nature, his soul is subtle, and his spirit or intellect is supra-formal. The human soul is an object of the subtle state that is endowed with consciousness; it thereby gives man his individual subjectivity. Because man has a body that is an object of a gross nature, he finds himself outwardly in a world of a gross nature that is our terrestrial world—of which he is the central being. Now, consciousness is not material; as such, conscious beings do not necessarily have to have gross bodies. If we find conscious beings in possession of bodies all around us in our terrestrial world, we should have no reason to deny the existence of analogous conscious beings who may not have gross bodies. Psychic beings[12] of all kinds inhabit the subtle state that envelops our terrestrial world; they are conscious beings in possession of souls and spirits, but are without bodies. The subtle state itself, we said, is enveloped by the supra-formal state, which is the realm of the angelic beings—beings of pure spirit or pure intelligence not charged with individual souls and gross bodies.[13] When appearing in the lower worlds, however, angels reflect themselves in these states, and thus take on forms that are akin to the beings of those worlds.

There is a creative rhythm emerging from the infinitude of the Principle. To say Principle is to say Creation. The Universe is, as it were, a divine "respiration." The degrees of real-

---

12 There is no people whose tradition does not include some beings of the subtle world. The nymphs of the Greeks, the elves of the Germans, the peris of the Persians, the jinn of the Arabs, and the rakshasas of the Indians are only a few examples of a very long list of varied types of psychic or subtle beings—beings that are well-accounted for in a cosmology that includes the subtle world in its scope. The phenomena involving subtle beings demonstrate fissures in the purely materialistic "cosmologies," and thus serve as evidences proving the inadequacy of their premises.

13 Sometimes "angels of fire" and "angels of light" are spoken of. "Angels of fire," then, represent a variety of the beings of the subtle state.

ity have "not only a static aspect or an aspect of simultaneity, but also a dynamic aspect or an aspect of succession, and this evokes the doctrine of the universal cycles: each of the degrees of the Universe contains a different cyclical rhythm, which means that this rhythm of apparition or manifestation—or of 'crystallization,' whether principial or existential, depending on the case—becomes more and more 'rapid' or 'multiple' as one moves further from the immutable Center; and this is expressed geometrically by the increase of surface towards the periphery. Therein lies the whole doctrine of the 'days,' 'years,' and 'lives' of *Brahmā*,"[14] of which the Hindu doctrine of the universal cycles gives a very explicit account.

### Symbolism: The Language of Metaphysics

Metaphysical truths—by nature surpassing the plane of the human mind—are not proven, and are not provable, by rational means. The language of metaphysics is of necessity descriptive and makes use of the things of our world to refer to the realities that transcend the latter. This explains why Plato, who so characteristically refutes false or inadequate doctrines by much use of reason, is often content to present metaphysical truths by use of a descriptive language and speaks in allegories and analogies.[15] A metaphysical doctrine will appear to the rationalist mind as nothing but dogmatic speculation. The symbolical description, however, is based on the analogy between the communicating symbol and the truth thereby communicated, and thus serves to actualize a knowledge that is not added from without but is virtually contained in intelligence itself.[16]

14 Frithjof Schuon, *Form and Substance in the Religions*, pp. 61-62.

15 The allegory of the cave and the analogy of the divided line, in his *Republic*, for example.

16 See Frithjof Schuon, *The Eye of the Heart* (Bloomington, IN: World Wisdom, 1997), p. xiv.

The things of the material world reflect realities of higher orders. In a sense, they *are* the higher realities themselves— though on the material plane. This explains the meaning of symbolism. A symbol[17] then is the reality it symbolizes as far as that reality can be manifested in this world, and is thus an effective vehicle for it.[18] Water, for example, by virtue of being the symbol, in the material world, of the Universal Substance, has the effective power of purification for a believer, that is, it returns accidents to the Substance.[19]

The earthly world is the cradle in which man finds himself *a priori*. It is in the nature of things, therefore, that he be guided to higher realities by way of the symbols in this world. This is not only unavoidable, given man's terrestrial condition, but also perfectly sufficient given the essential identity of the things "below" with the things "above." That is why all the great traditions make use of human language and earthly symbols to recall to man the truths of higher realities. Sacred Scripture is full of images taken from our material world that are intelligible to all men, which is why the prophets and the saints speak a language that is, in principle, comprehensible to all men. Angels, who are beings of a "higher" order, when appearing on earth, appear to come from above, from heaven, as countless testimonies of the saints in all traditions bear witness. The "higher" and "lower," the "above" and "below," of our world symbolize, effectively and sufficiently, the hierar-

---

17 The word "symbol," from the Greek σύμβολον, "throw together," is a token of identity: it originally meant half of a broken object, which, when fitted to the other half, verified the bearer's identity. A symbol is far from being a mere conventional representation; it is, by virtue of an essential identity, the thing it symbolizes—on a lower plane, of course.

18 Sacraments are symbols become dynamic. The spiritual efficacy of sacraments is thus related to the analogy between the symbol and the reality it symbolizes. Moreover, because they are divinely ordained, sacraments convey God's presence in a direct and efficacious way.

19 See Frithjof Schuon, *Roots of the Human Condition* (Bloomington, IN: World Wisdom, [1991] 2002), p. 58.

chy of the successive worlds, precisely because these vertical relationships reflect themselves accordingly on the horizontal plane that is our terrestrial world.

The phenomena of the terrestrial world represent God's creative intention most directly by their appearance. Appearance, therefore, takes precedence over mechanism. That the sun rises in the east and sets in the west is not an arbitrary contingency; it directly manifests the Creator's intention and provides man with what he needs spiritually. Otherwise, mankind would have had to wait for thousands of years for a Copernicus to discover the heliocentric "truth." The Ptolemaic image of the world is how things appear naturally to man. Moreover, this image provides a spiritually efficacious cosmic symbolism: man is situated at the center of the universe, enveloped first by the sublunary world (comprising the four elements earth, water, fire, and air) that corresponds to the material world, then by the seven planetary spheres that correspond to the sensorial world, then by the sphere of the fixed stars that corresponds to the subtle state, all enclosed within the Empyrean that corresponds to the supra-formal state or celestial Manifestation.

The heliocentric image of the world too can have a symbolical meaning: the sun would then reflect the Principle in our material world. This is, however, beside the point, for what falls within man's normal experience, as we have said, takes precedence over what may be observed under very particular circumstances, for man's normal experience is universally accessible to all men throughout all the ages—and is thus more directly reflective of the creative intention of God. In Schuon's words:

> When God created man in His image, He created a measure; the human perception of the world corresponds to God's creative intention. . . . There are two points to consider in created things, namely the empirical appearance and the mechanism; now the appearance manifests the divine intention . . . the

mechanism merely operates the mode of manifestation. For example, in man's body the divine intention is expressed by its form, its deiformity, its symbolism, and its beauty; the mechanism is its anatomy and vital functioning. The modern mentality, having always a scientific and "iconoclastic" tendency, tends to over-accentuate the mechanism to the detriment of the creative intention, and does so on all levels, psychological as well as physical; the result is a jaded and "demystified" mentality that is no longer "impressed" by anything. By forgetting the divine intention—which nonetheless is apparent *a priori*—one ends in an emptiness devoid of all reference points and meaning, and in a mentality of nihilism and despair, if not of careless and brutal materialism. In the face of this deviation it is the child who is right when he believes that the blue sky above us is Paradise.[20]

## The Modes of Reality

There are not only degrees in reality, there are also modes; the former are in "vertical" order while the latter are "horizontal" and situated in the appropriate manner at each degree.[21] Modalities are best described by the Pythagorean numbers, which are essentially qualitative notions that are descriptive of the structure of reality.[22]

---

20 Frithjof Schuon, *Roots of the Human Condition*, pp. 17-18.

21 See Frithjof Schuon, *To Have a Center* (Bloomington, IN: World Wisdom, 1990), p. 142.

22 Number, in itself quantitative, has something qualitative when it is abstract. Unlike ordinary numbers that are obtained by addition, numbers in the Pythagorean sense result from an intrinsic differentiation of the principial unity. Geometrical figures are so many images of unity: for example, the triangle represents harmony, and the square, stability (see Frithjof Schuon, *Gnosis: Divine Wisdom* [Bloomington, IN: World Wisdom, 2006], p. 93).

From the perspective of unity, everything owes its being and suchness to the Supreme Principle. Modalities, however, only begin to unfold starting with duality. The Supreme Principle is absolute. It admits of no determination, hence no limitation; it is thus infinite. The Absolute excludes all that is not it; it accepts no other. The Infinite includes all; nothing remains outside it. The duality Absolute-Infinite appears as two complements in the human mind, but is undifferentiated in the Essence. This undifferentiated duality is the foundation of all dualities: the active or masculine pole and passive or feminine pole—in all orders, starting from that of the Divine.[23] The Divine Qualities of Rigor and Gentleness refer to the same duality.[24] And this duality reverberates throughout all existence in a differentiated mode.[25] Man is of necessity manifested in two forms.[26] The duality Absolute-Infinite is, as it were, the warp and weft of all creation. In space, the Absolute manifests itself in the point, and the Infinite in extension. In time, the Absolute manifests itself in the instant, and the Infinite in duration. In the human body, the Absolute manifests itself in its skeletal structure, the Infinite in the indefinite diversity of its forms. In language, the Absolute manifests itself in grammar, and the Infinite in the indefinitely varied contents. In any definition, the Absolute manifests itself in exclusivity, and the Infinite in inclusivity. One could proceed indefinitely: for example, breathing consists of two phases, inhalation and exhalation, and the heartbeat, of two cycles, contraction and expansion—reflecting, correspondingly, the Absolute and the Infinite.

23 *Purusha* and *Prakriti* in Hindu terms, for example.

24 Qualities of *jalāl* and *jamāl* in Islamic terms, for instance.

25 "And of all things We created pairs, that ye may reflect" (Koran 51:49).

26 "So God created man in his own image, in the image of God created he him; male and female created he them" (Gen. 1:27).

From the perspective of trinity, the Supreme Principle may be envisaged, in itself and without differentiation, by the triad Being or Reality, Consciousness or Knowledge, and Beatitude or Felicity: *Sat, Chit, Ānanda* in Sanskrit, or *Wujūd, Shuhūd, Sa'ādah*, in Arabic. This triad reflects itself in all degrees, including that of Being. The Christian Trinity too, in its horizontal relationship, can be viewed as an expression of this wherein the three Persons are the three hypostatic functions of Being. The perspective of trinity is likewise reflected in all existence, first giving rise to the numerous deities of the mythological religions and ultimately reflecting itself in man, giving him his ternary aspect: will, intelligence (or thought), and love (or sentiment in its most profound sense).

The perspective of quaternity may be described by four aspects, namely, purity, life, strength, and peace, which are the origin of all qualities, divine as well as cosmic. In the Divine Order, they correspond to fundamental Divine Qualities, which are reflected in all of existence. The four fundamental cosmic qualities for the American Indians correspond to the four cardinal points: north, south, east, and west. In the microcosm, the transpersonal intellect reflected in the human mind, refracts, as it were, into four principal mental faculties of reason, intuition, imagination, and memory. Quaternity, moreover, opens onto indefinite differentiation, which reverberates throughout all creation.

### Consequences of the Lack of a Metaphysical Perspective

In the absence of an integral metaphysical perspective that includes all degrees of being, much remains inexplicable. A doctrine of being that ignores the higher degrees of reality shows itself inadequate in the face of the innumerable phenomena—inside as well as outside man—that it cannot explain, and if it integrates such an ignorance of higher realities in its principles, it cannot but lead to logical inconsistency.

The phenomena of man's own subjectivity—his consciousness, his access to eternal ideas or principial realities, hence his objective intelligence, for instance—as well as the phenomena of the objective world—the very existence of things and the immateriality of life, for example, not to mention the psychic and spiritual phenomena that cannot be explained solely by reference to the lower degrees of being—all provide evidences pointing to the inadequacy of the doctrines that fall short of accounting for all degrees of reality. Moreover, such restrictive perspectives are intrinsically inconsistent in their axioms since, characteristically, the very thesis that postulates a restriction in such perspectives does not satisfy its own postulated criterion, and thus contradicts itself. For example, the overarching principle of materialism—that only matter is real—is not itself material, and, as such, it contradicts this very thesis; it is, therefore, intrinsically inconsistent. Likewise, the guiding principle for many schools of modern psychology—for which reality is extended to the realm of the psyche, but not beyond it—presents itself as the solely "objective" verdict on the "subjective" nature of the human soul. It, therefore, contradicts itself in its very principle.

Metaphysics provides fundamental keys to the understanding of the structure of reality; it also provides the vantage point that man needs to be able to position himself in the universe. In the absence of these keys, serious errors are inevitable. When modern man, who, by and large, is ignorant of these fundamental keys, encounters the phenomena that show fissures in his worldview, he often resorts to denying their existence, or sets them aside as things that science will unlock in the future, or else falls prey to false speculations of a "spiritualistic" nature, which in reality are nothing more than formulations of superficially psychic perceptions.

Metaphysics is not rationally provable; nor is materialism, of course. Materialism cannot prove the existence of matter, nor can it prove the inexistence of realities beyond matter. It merely takes the exclusive reality of matter as an axiom. A thorny question for the materialist is that of the nature of consciousness—a clearly immaterial reality—which, for want of a better explanation, he tries vaguely to attribute to the processes in the human brain and nervous system, for example; or worse still, given that he himself finds such vague explanations unconvincing, he leaves it out as a question for science to explain in the future. Consciousness, then, miraculously jumps out of lifeless matter. It would thus be an effect of matter, an effect that would then have command of its cause, and, as such, would be superior to it, which is absurd.[27] Intelligence would be purely material, its operation purely mechanistic and rooted in the biochemistry of cells. One wonders, then, what difference there is between minerals and living beings, and further between vegetables and animals, and ultimately between animals and human beings.

The worldview of the materialist shatters when he encounters the phenomena of orders higher than that of matter; assuming, of course, that he does not readily deny their existence. A physicist, for example, sees the cause of all movement in physical forces, and according to the laws of physics that he knows; his worldview collapses when he sees objects in motion without an apparent physical cause as in a spiritist séance, for example, or when he encounters the wonders a sorcerer can work. Alas, he will then be inclined to believe in, not the epistemological and metaphysical doctrines of the religions—which present a comprehensive account of reality—but whatever explanation he might receive from a spiritist or a sorcerer—who is merely setting into motion the forces, often of a dark nature, of the subtle or psychic state, and whose idea of spirituality is as far removed from truth as that of the materialist, if not further. In this sense, then, one

---

27 The greater cannot come from the lesser; otherwise, the differential would be an effect without a cause, thus contradicting the law of causality, which is the foundation of logical thinking.

might say that materialism prepares the way for "spiritualism," or occultism, in its many facets.

From the point of view of metaphysics, the subtle state is only a degree of the natural world—astonishing as this may sound to the materialist—and, as such, in itself, it has nothing supernatural or spiritual about it. The world of the spiritual, properly so called, is the angelic world, which is out of reach for the psychic worker.[28] Here the knowledge of the supernatural, that is to say, that of the degrees beyond the subtle state, becomes crucial, and that is why only a full doctrine of the nature of reality can help put things in their proper place. In fact, seeking exposure to the subtle world—communion with spirits, witchcraft, necromancy, and psychic experiences of every kind assisted by "meditation" or induced by the use of psychedelics, to name only a few in this genre—can be more harmful than pure and simple materialism, since man would thereby render himself the passive recipient of dark psychic forces. And this explains why the orthodox religions condemn such practices in the strongest possible terms: practices that are ubiquitous in our post-modern world, driven as they are by a "new age" kind of "spiritualism."

In this order of ideas, one may refer to so-called near-death experiences (NDEs), which for many of our contemporaries—practically materialistic or at most psychic in their worldviews, hence unaware of the "discerning of spirits"[29]—have become the source of eschatological doctrines. Although it is exceptionally possible for a deceased person to return to life by miracle with a message of the afterlife—the example of Er mentioned by Plato in his *Republic* shows the possibility—it is hard to imagine a divine intervention for innumer-

28 Although angels may reflect themselves in the formal—subtle as well as gross—state when they intervene in it, they are still not at the psychic worker's disposal.

29 "Beloved, believe not every spirit, but try the spirits whether they are of God: because many false prophets are gone out into the world" (1 John 4:1).

able people in our time,[30] only to preach a universal message of unconditional love[31] that requires nothing of man but the abdication of all discernment towards every doctrine and every man, good or bad, promising heaven to everyone,[32] and effectively rendering the great religions superfluous.

Here we do not wish to cast doubt on the veracity of those who have had these experiences and have come back with such messages, which often instill in them concrete convictions about the immortality of the soul and the immateri-

---

30 The interference of psychic beings in the affairs of men is nothing new. Legends of the traditional worlds are full of such things. The proliferation of those NDEs that promise a happy afterlife to everyone, especially in the modern world where these phenomena and the messages they bring find fertile ground, may be due to the free-hand such beings are given in an ambience where the protective measures that the great religions make available to their followers through sacraments and rituals are further and further weakened because of man's accelerating loss of faith in the orthodox religions. Furthermore, psychic and "spiritualistic" practices, fashionable in ubiquitous occultist and "new age" circles, could help pave the way for such interferences.

31 The commonplace message of an unconditional love that promotes the idea of loving all men as they are without the slightest concern about their errors is an abuse of the notion of charity. Christ's second great commandment is a function of his first: "Thou shalt love the Lord thy God with all thy heart, and with all thy soul, and with all thy mind. This is the first and great commandment. And the second is like unto it, Thou shalt love thy neighbor as thyself" (Matt. 22:37-39). The first commandment is unconditional, the second is not. Christ could not have meant that one should love one's own errors, or those of others for that matter, else why would he have said, "If any man come to me, and hate not his father, and mother, and wife, and children, and brethren, and sisters, yea, and his own soul also, he cannot be my disciple" (Luke 14:26)?

32 Some psychologists, and especially psychoanalysts, who in our time have practically usurped the place of the spiritual counselors of the traditional worlds, ever-intent on ridding people of complexes, have become the pioneers of the "NDE science," and wish to rid people, even the dying, of their "complex" of fear of death by the happy message of the unconditional, everlasting bliss that awaits all the deceased, thus eliminating the last possible chance for conversion and redemption.

ality of consciousness. What we contest is the nature[33] of such experiences and the source of such messages. These extraordinary experiences are impressive to modern man, precisely because his worldview is conditioned by materialism to one degree or another. If he were anchored in a more comprehensive view of the world, he would be better equipped at dealing with extra-sensorial experiences, as our ancestors generally were, those for whom "personal experience" in itself was not so much a source of truth as was their sense of the Absolute, their faith in God, and their intellectual convictions. Modern man's worldview, even that of a person nominally and superficially within the framework of an orthodox religion, hardly has any room for psychic phenomena; this renders him susceptible to false teachings from doubtful sources. In an authentic spirituality, the spiritual "wayfarer" is advised not to seek "experiences," and even if he has one spontaneously, he will subject it to intellectual and scriptural criteria, to his spiritual advisor if possible, in stark contrast to modern "spiritualism" where "personal experience" tends to become everything.

Let us return to the errors that are inevitable in the absence of a proper metaphysical perspective. As we move up in the hierarchy of the degrees of being, to the Divine Order, things appear more subtle to the human mind, yet they maintain their importance. The necessity of a correct understanding of the degrees of the Divine Order, for example, becomes clear when one looks at the misunderstandings our contemporaries, especially in the West, have of Buddhism, a religion that has no concept of a Personal God. *Nirvāna* or the Void, in this case, refers to the Divine Essence, and as such, Buddhism, in its doctrine, like all other orthodox religions, has the Absolute in view. Buddhism, therefore, is not atheistic,

---

33 There may indeed be no "death" in NDEs. Rather, the body's grip on the soul is loosened by an external event, and the soul may merely be exposed to the subtle world, to which it belongs by nature.

as some would like to think, but may be termed non-theistic to account for its lack of an explicit reference to a Personal God. This character of Buddhism has ironically rendered it appealing to some, especially in the West, who are repelled by a Law-giving Personal God, and the commitments on the part of man that His reality implies. But this is simply a misconception due to their lack of a proper metaphysical perspective.

Metaphysics, we see, is fundamental to a correct understanding of the nature of reality, and is thus fundamental to a correct understanding of the nature of science in the true sense of the word.

# CHAPTER 4

# THE DOCTRINE OF CAUSES

Science, for Aristotle, is the knowledge of causes: "We think we have knowledge of a thing only when we have grasped its cause."[1] Accordingly, in order to grasp the notion of science, we will have to develop an understanding of causes.

Aristotle identified four causes: (i) final, (ii) efficient, (iii) formal, and (iv) material, in response to questions of the (i) purpose, (ii) agent, (iii) design, and (iv) material of a thing. For example, the purpose, or final cause, of a wooden dining table is to dine at it; its agent, or efficient cause, is the carpenter who made it; its design, or formal cause, is its shape; and its material, or material cause, is wood.

Aristotle's doctrine of causes is incomplete in itself, as we shall see when we consider the comprehensive account of the Platonic doctrine of causes. Even so, since the time of Galileo and Newton, the general tendency of scientists has been to ignore the final cause, and to treat natural phenomena from a more limited perspective—although, more recently, there have been voices that call for the reinstatement of the final cause into the scientific perspective.[2] The modern scientist, by and large, is concerned with the mechanism of creation. As such, modern science is the knowledge of the lower Aristotelian causes.

---

1 *Posterior Analytics*, 71 b 9-11. See also *Posterior Analytics*, 94 a 20 and *Physics*, 194 b 17-20.

2 See, for example, the articles in *Newton to Aristotle: Toward a Theory of Models for Living Systems*, edited by John Casti and Anders Karlqvist (Boston, MA: Birkhäuser, 1989).

It has been said that Plato's philosophy is heavenly, but that Aristotle made it earthly. According to Marsilio Ficino, "Plato deals with natural things divinely, while Aristotle treats divine things naturally."[3] In other words, Aristotle's philosophy is Plato's philosophy become earthly. And in the words of St. Bonaventure, "It seems that, among philosophers, it was given to Plato to speak of wisdom, to Aristotle of science. The former looked mainly towards the higher things, the latter mainly towards the lower."[4] This can be seen clearly in their respective doctrines of causes. For Plato, there are six causes: (i) final, (ii) paradigmatic, (iii) efficient, (iv) instrumental, (v) formal, and (vi) material.[5] The paradigmatic cause is the model after which a thing is made, and the instrumental cause is the instrument by which a thing is made.

Plato refers to the first three causes, namely, the final, paradigmatic, and efficient causes, in the following passage from his *Timaeus*:

> It is, moreover, a question of knowing after which of the two models the Creator and Father of the world constructed it: after that which is immutable and always the same, or after that which has come into existence. If the world is beautiful and its Creator good, undoubtedly, He constructed it after the eternal model . . . for the world is the fairest of all that has come into existence, and its Creator the best of all causes. . . . Let us state the reason the Creator

3 Quoted in R.E. Proctor, "Beauty as Symmetry, Part II," *American Arts Quarterly* (2010), Vol. 27, No. 2.

4 *Sermones selecti de rebus theologicis*, "Sermo IV: Christus unus ominum magister," 18.

5 The origin of the doctrine is Plato, but its analytical presentation comes to us from the Neoplatonists. For an elucidation of Plato's six causes, refer to the second book of *The Commentaries of Proclus on the Timaeus of Plato in Five Books*.

had in forming the becoming and the world. He was good, and he that is good never has envy concerning anything, and being devoid of envy, He desired that all should be, so far as possible, like unto Himself. Whoever, instructed by the wise men, accepts this as the main reason for the becoming and the world shall be wholly right.[6]

The efficient cause of creation is the "Creator" who is "the best of all causes."[7] The paradigmatic cause is the "model" after which the Creator constructed the world—a model that is "eternal." This is an important cause that Aristotle left out of consideration, and to which we shall soon return. The Creator, Plato says, "was good," and, as such, "desired that all should be, so far as possible, like unto Himself." This explains the final cause of creation, that is, to manifest the Sovereign Good,[8] which prefigures the Augustinian notion that it is in the nature of the Good to wish to communicate itself. The Platonic notion of the final cause of creation is in conformity with the teachings of the great religions.[9]

## The Instrumental Cause

In his account of cosmology in the *Timaeus*, Plato offers mathematical descriptions of the structure of the world, where we see that mathematical proportions and relations—arithme-

6 Plato, *Timaeus*, 29a-29e.

7 "The best of creators," to use a Koranic phrase (23:14).

8 Plato, when referring to the Divine Essence, uses two terms: "the One" in his *Parmenides*, and "the Good" in his *Republic*, Book VI, and also in his *Philebus*. In this context, reference is made to Proclus' *Theology of Plato*, Book II, Chapter 6.

9 Witness, for example, this Divine utterance from the Islamic tradition: "I was a hidden treasure and I wished to be known; therefore, I created the world."

tic, geometric, and harmonic means, for instance—are said to have been used in the creation of the bodily and animic worlds.[10] According to Proclus' commentary, the mathematical proportions and relations that govern a system constitute its instrumental cause. According to him, this was also the perspective of the Pythagoreans on mathematics and its role in the structure of the world.

Thus, for example, the same proportions describe the mathematical relationships among the frequencies of the musical tones in the realm of sounds, and the colors of the rainbow in the realm of visible light. And these same proportions are related to those of the planetary distances, whence the concept of the music of the spheres, which, it is said, Pythagoras could hear, not by his bodily ears, of course, but deep in his heart.[11] In Rumi's words:

> The shrill of the clarion and the menace (beating) of the
>       drum
> (they) are a little like the Universal Trumpet.
> Hence the wise have said that these harmonies
> we received from the revolution of the (celestial) spheres.
> It is the sound of the revolutions of the sphere
> this (melody) that people sing with pandura and throat.[12]

What is objectively in accordance with these laws appears subjectively pleasing to man, precisely because these same objective and universal laws are built into man's perception, or because man, with these realities inscribed in his spirit, is able to recognize and appreciate them in the objective world outside him.

10 In the language of the Koran: "Verily, We have created every thing by measure" (54:49).

11 It is thus that "the seven heavens and the earth and all that is therein praise Him, and there is not a thing that hymneth not His praise; but ye understand not their praise" (Koran 17:44).

12 *Mathnawī*, Book IV, 740-742.

The same mathematical structure may describe different physical systems, but it remains the instrumental cause of each and every one of them.[13] Mathematical relations that are inherent in creation, and which render it beautiful and harmonious, are thus its instrumental cause.

Much of modern science, particularly in mathematical physics, has to do with the discovery of the mathematical structures that govern physical systems, hence with their instrumental causes. These efforts would be considered science in the Aristotelian sense; nonetheless, they belong to the lower end of the spectrum of science, taken as a whole, given that they solely aim at the discovery of mechanisms. For traditional man, on the other hand, "the nobility of a science is of the same measure as the nobility of its subject matter,"[14] and as such "the nobility of a science depends on the certitude it establishes."[15] Accordingly, for him, a science worthy of pursuit was one that looked up towards the higher causes, namely, towards that which defined man's reason for being.

## The Paradigmatic Cause

Returning to the paradigmatic cause, Plato goes on to explain it in more detail in his *Timaeus*:

> In the likeness of which living being did the Creator of the world compose it? Certainly, it could not have been in the likeness of those which belong by nature to the category of parts; for that which resembles the

13 For example, conventionally, simple vibration is described by a second-order linear differential equation. Whether the physical system is electrical or mechanical, its governing mathematical equation, that is to say, its instrumental cause, remains the same.

14 According to al-Ghazali, an eminent eleventh-century Sufi, philosopher, and theologian.

15 According to St. Thomas Aquinas (*Summa Theologiae*, I, Q. 1, Art. 5, ad. 1).

imperfect cannot be beautiful. We shall affirm, then, that the world, more than anything else, resembles that of which all other living beings, individually and generically, are parts. For that contains within itself all the intelligible living beings, just as this world contains us and all other visible animals. For since God desired to make the world resemble most closely that which is the fairest and most perfect of all intelligible living beings, He constructed it as a living being, one and visible, containing within itself all the living beings that are by nature like itself.[16]

The "intelligible living beings" are the Platonic ideas or archetypes. They are the principial realities that lie at the foundation of every being, in its potentiality or in its existentiated form.[17] They are the models upon which things or beings of this world are patterned, whence the term paradigmatic cause.

As will be discussed extensively in the next chapter, in the language of the Bible and the Koran, these models are the "names," the knowledge of which Adam had within himself thanks to his deiformity. In seeing every being or thing, Adam saw its archetype, that is, the Divine Name of which that being or thing was a manifestation, and to which it returns in its essence. For primordial man, the phenomena of this world had metaphysical transparency: they conveyed to him their archetypal messages. Accordingly, to know a thing is not to merely take account of its outward appearance or properties, as modern science does, but to see its metaphysical transparency, the Divine message that lies at its root. The thing, then, becomes a symbol of the archetype it symbolizes; it becomes

---

16 Plato, *Timaeus*, 30c-30d.

17 In the language of the Koran, the "Pen" writes out the cosmic possibilities upon the "Tablet." In every phenomenon and at every level, there is an "idea" which is incarnated in an existential receptacle (see Frithjof Schuon, *Esoterism as Principle and as Way* [Middlesex, UK: Perennial Books, 1981], p. 51).

a "sign" of it. There is thus an essential identity between the symbol and the reality it symbolizes. In this sense, therefore, things of this world become signs of God.[18]

The knowledge of symbols and symbolism is a science in the true sense of the term, and a noble one, since it has the higher causes in view. It is, however, a science that has practically been forgotten by the "thinker" of the past few centuries, who no longer concerns himself with what is "above," but is content merely with the things here "below," uprooted as they are from their higher causes.[19] Here one is reminded of this Zen metaphor: "When the finger points to the moon, the foolish man looks at the finger."

Time is a universal container that encompasses everything in the world. Its paradigmatic cause, for example, is beautifully described in the same dialogue as God's eternity:

> When the Father Creator saw this moving and living image of eternal gods, he rejoiced, and in his joy resolved to make it still more like its model; and as that model was an eternal living being, he sought to make the world in like manner, so far as it might be possible. But the nature of that living model was eternal, and thus its full adaptation to a thing that comes into existence was impossible. Wherefore he resolved to have a moving image of eternity; and by

18 It is also in this sense that Shabistari says, "If the Muslim but knew what an idol is / he would know that faith is in idol worship" (*The Garden of Mystery*, 872).

19 It was René Guénon's works, "the voice of him that crieth in the wilderness" of the twentieth-century West, that called out against the modern world and spoke of the importance of symbolism in the context of metaphysical doctrine and traditional principles (see especially his works *The Crisis of the Modern World* [Hillsdale, NY: Sophia Perennis, (1927) 2004], *The Reign of Quantity and the Signs of the Times* [Hillsdale, NY: Sophia Perennis, (1945) 2004], and *Fundamental Symbols: The Universal Language of Sacred Science* [Cambridge, UK: Quinta Essentia, (1962) 1995]).

the order he set in heaven, he made of eternity, which rests in unity, this image, eternal but moving according to number, which we call time.[20]

Time, then, is a moving image of God's eternity, and, as such, never stops; rather, it comprises cycles, hence a rhythm.[21]

## The Importance of the Paradigmatic Cause

On the necessity of taking account of the paradigmatic cause of things, suffice it to say that in the absence of this principle, the fundamental link between the world and its Cause is broken;[22] what remains is at most an abstract notion of God as the Creator. This, in fact, renders man prone to the Deistic error, according to which God created this world with apportioned laws that spare Him His presence and intervention in the world. Man can then investigate the things of this world with no need for God, since his knowledge of phenomena is of no use to his knowledge of God. The things of this world are no longer "signs," and the world is deprived of its metaphysical transparency. The stage, then, is set for forgetting the Creator altogether, or even for denying Him. The final cause of creation is no longer the manifestation of the Sovereign Good, but the remote ontological principle of Aristotle.

20 Plato, *Timaeus*, 37c-37d.

21 Whence the doctrine of the universal cycles, to which allusion was made in the preceding chapter.

22 In this context, it is worth noting that the rise of nominalism during the late Middle Ages—especially in the form of the rejection of the Platonic realism that sees concrete realities or "ideas" in the universals—paved the way for the subsequent errors that found fertile ground in the post-Renaissance West. The nominalist fideism of a William of Ockham, for example, saw in God only a remote ontological necessity; science, then, was a matter of exploration and discovery, no longer connected to the metaphysical principles that were formerly its foundation. Science was thus left to its own devices, soon to "lose its own soul" in the process, in trying to "gain the whole world."

Symbolism, on the other hand, finds its foundation in the paradigmatic cause. The phenomena of this world are symbols of their divine archetypes, and can thus serve to lead us back to their source. The symbolist language of sacred Scripture will then be understood, and descriptions of higher realms will thereby become intelligible. Paradisiac scenes in sacred Scripture are certainly not to be taken literally, as we know, but they would be vague, propagandistic means were they mere metaphors. They are, on the contrary, exact doctrines once they are understood through their paradigmatic cause, that is, in the light of the science of symbols, in the true sense of the term.

Witness this Koranic imagery, for example:

> The symbol of the Garden that the righteous are promised: therein are rivers of water unpolluted, rivers of milk whereof the flavor changeth not, rivers of wine that are a delight to the drinkers, and rivers of clear-run honey; therein for them is every kind of fruit....[23]

The meaning will be clear when it is remembered that "that which is below is like that which is above and that which is above is like that which is below."[24] For instance, the Divine Name of which milk is an earthly manifestation, appears in Paradise as pure of all earthly contingencies; that is to say, it is a paradisal milk of which the "flavor changeth not," meaning that in God's proximity it is no longer a veil upon its archetype.

Symbolism, we said, is an "exact" science. Now, in conformity with the Platonic principle that like attracts like, Plotinus states that:

---

23 Koran 47:15.

24 According to the *Emerald Tablet*, which is thought to be the work of Hermes Trismegistus (the quoted translation is Isaac Newton's).

It is always easy to attract the Universal Soul . . .
by constructing an object capable of undergoing its
influence and receiving its participation. The faith-
ful representation of a thing, thus, is always capable
of undergoing the influence of its model; it is like a
mirror which is capable of grasping the thing's appearance.[25]

For an archetype to be present in its earthly symbol, a
sanctuary or an icon, for instance, it is important that the
symbol, by its proper form, adequately represent the reality
it is meant to symbolize; it therefore has to conform to universal laws that exclude all human arbitrariness. This explains
why sacred art for traditional man was an "exact" science that
required a profound understanding of universal truths as well
as a strict preparation and training.[26]

Ancient languages, unlike their corroded descendants
that are our modern languages, provided, by their formal
richness, distinctions that accounted explicitly for fundamental causes.[27] The declension of a noun, then, conveyed
substantial information about its function. In Sanskrit, for
example, the final cause can be expressed by use of the dative
case; the efficient cause is related to the nominative case, and
the instrumental cause, to the instrumental case; the ablative
case is used to express the origin and also the material cause.

In this order of ideas, it is worth noting that fundamental
words in ancient languages were still akin to their primordial prototypes; they were thus symbols of the realities they

---

25 *The Enneads*, IV.3.11.

26 One is referred to Titus Burckhardt's *Sacred Art in East and West: Its
Principles and Methods* (Middlesex, UK: Perennial Books, 1967), for an ex-
cellent treatment of this subject.

27 Which, let it be noted in passing, is indicative of the sense of discernment the speakers of such languages must have had. It is hard to imagine
that the corrosion of languages over time constitutes an "evolution."

manifested in the domain of sounds; as such, they echoed the creative divine words that were the principles of the things of this world.[28] In this sense, their very form, or sound, was reflective of their paradigmatic cause. In general, sacred Scripture, by virtue of being God's word, vehicles God's presence in the most direct manner in the domain of language. Given its theurgic power, its recitation is not merely a mental assimilation of ideas, but an existential participation in the Divine Presence.

Modern man, by and large, lacks a full understanding of causes, especially the higher ones, namely, the final and paradigmatic causes. Modern science is solely preoccupied with the mechanism of things, never with their higher causes; but sometimes, by a misuse of language, it presents formulations that pass for explanations of such causes. The theory of evolution is a case in point. As a hypothesis, it may be used to rationalize and classify certain observations, but it cannot function as a substitute for an explanation of higher causes. It may be argued that variation and natural selection could be useful, particularly in biology, in explaining certain developments within species. Even so, and while admitting that very partial evolutions or superficial adaptations to given environments are always possible, the theory of evolution radically falls short of explaining those phenomena that surpass the scope of empirical investigations—the phenomenon of human subjectivity, for example. An acceptable theory, we have said, should be able to adequately account for the things that fall within its scope. Transformist evolutionism clearly fails this test by its glaring inability to explain such things as the emergence of life, intelligence, or consciousness from lifeless

28 In Hinduism, for example, a *bija mantra* is a one-syllable word having no literal meaning, which, when uttered, would invoke the presence of the deity it symbolizes. Perhaps not unconnected with the notion of the *bija mantra* is the mysterious case of the "disjoined letters" that open a quarter of the chapters of the Koran.

matter: it lacks any explanation about this supposed transition from the lesser (matter, in this case) to the greater (human intelligence, for instance). As regards the latter, the ancients would apply the principle that the greater could never come from the lesser, and as such would refute evolutionism at its root. Evolutionism, then, in denying the vertical dimension, posits hypothetical causes on the horizontal plane. In doing so, it ignores an entire dimension of reality, namely, that of the archetypes, and their penetration into our world through supra-formal and subtle formal states; in Schuon's words, "it is as if one wished to make a fabric of the wefts only, omitting the warps."[29]

Another error prevalent in the modern outlook is the confusion of causes. When physical laws in a mathematical language are discovered, for example, it is as if the whole mystery of creation were supposedly unlocked, whereas in reality these laws are merely explanations of the lower causes. But this does not matter to the modern scientist since the scientific outlook has long left the troubling "why" of creation behind, and blithely occupies itself with its "how."

---

29 Frithjof Schuon, *To Have a Center*, p. 51.

# CHAPTER 5

# SCIENCE IN THE MIRROR OF THE RELIGIONS

No traditional people ever lived without a religion, which bound them to the Absolute, defined their reason for being, and fashioned them in every respect. To speak of the great traditions that shaped humankind throughout the ages—excepting recent history when man has considered himself emancipated from the bonds of religion—is thus to speak of the religions.

The scope of modern science has been deliberately restricted to the empirical sciences; this tacit assumption, however, has in practice been forgotten—or its importance underestimated. Science for traditional man had a much wider scope; it encompassed not only subjects of practical importance such as medicine and agriculture, but also disciplines such as astrology and alchemy. Things alien to the modern scientific mentality, such as magic, were also considered science, though not necessarily of a good kind. Whereas the practical sciences took account of man's earthly existence, the sacred sciences had man's final ends in view.[1] Traditional cosmology, ontology, and, ultimately, metaphysics were forms of science that dealt with objective reality while man's subjective reality was taken into account in disciplines such as epistemology, or on another plane, in traditional psychology, the science of man's soul.

---

1 "Science is of two kinds," according to an Islamic teaching, "the science of the religions and the science of the bodies," that is, those sciences that have the vertical dimension in view and those that take account of man's horizontal existence.

For the men of wisdom in all the great traditions, the search for truth invariably started with the consciousness of the Absolute—the foundation of all truth and certitude—and thus with the discernment between the Real and the illusory, and the repercussions of this discernment throughout all the degrees of reality. Traditional man knew that there is no end to the knowledge of phenomena. His goal was not to collect and archive facts, but to seek the truth. He was wary of losing himself horizontally; he gave priority to surpassing himself vertically.

Before the Age of Reason, man's faculties of knowledge were not deemed limited to the sensory and rational faculties, and were thus commensurate with the likewise unrestricted object of knowledge that includes realities or beings beyond the realm of matter. Therefore, to speak of supra-sensorial or immaterial beings—fairies or angels, for example—was as natural to the ancients as was the truth that man's reason receives its light from something both higher and deeper in the human spirit.

In our search for what science meant to the ancients, and to avoid limiting its scope, we will need to be as comprehensive as possible, encompassing all that was called science or knowledge, namely all things in which we discern extrinsically a person who imparts knowledge and another who receives it, as well as an object of transmission that is knowledge itself. This object should have an objectively discernable value; otherwise, its existence, or lack of it, would be a matter of indifference.

For traditional man, the great imparter of knowledge was first and foremost God, often through revealed Scripture, then the prophets, sages, and saints. Traditional man seldom exaggerated the role of the sciences of the world, and yet, given his integral view of science, he would, in principle, not exclude the sciences of the world from the realm of science.

Sacred Scripture, we said, was often the most authoritative source of truth for the ancients. Accordingly, to gain an understanding of what science encompassed for traditional man, a look at a few examples in sacred Scripture is in order.

These examples show the great disparity between the outlook of the ancients and the sensibility of the scientific man of our time. It will be seen that things as unfamiliar as the interpretation of dreams, magic, and the language of birds were considered science according to the authority of sacred Scripture.

Here we should note that the defining characteristics of traditional man are vastly different from how modern man views himself. To be able to understand the way the ancients saw themselves in the universe, contemporary man needs a leap of the imagination, from his perception of a mechanical universe to one that is alive and teeming with consciousness and conscious beings. Thus, the reader is advised to suspend judgment for the time being and admit for the sake of argument that the following accounts in Scripture relate to reality, however fantastic they may appear to his imagination, conditioned as it has been by a materialistic worldview. In short, he should be open to the idea that, in Shakespeare's words, "there are more things in heaven and earth than are dreamt of in your philosophy."

## Examples of Science in Sacred Scripture

The story of Joseph, and in particular his knowledge of dreams, as related in the Bible and the Koran,[2] offers an ex-

---

2 Gen. 40-41 and Koran 12: A victim of his brothers' jealousy, Joseph was sold as a slave in Egypt, where he found favor with his master. Later, falsely accused of wrong-doing, he was put into jail. In all this, because of his righteousness, God was with him. While in prison, he successfully interpreted the dreams of two inmates, Pharaoh's former butler and baker. The butler, restored to service as Joseph had predicted, remembered him when Pharaoh began to have disturbing dreams. Joseph was brought to the palace to interpret Pharaoh's dreams, which Joseph interpreted to mean that there would come seven years of great plenty throughout the land to be followed by seven years of famine. Impressed by Joseph's knowledge and character, Pharaoh appointed him to a high office to oversee the management of the imminent crisis. Thus empowered, Joseph successfully led Egypt through the seven years of famine by having sufficiently stored up grain during the seven years of abundance.

ample of a science that, foreign as it may be to the modern scientistic mentality, was perfectly comprehensible to traditional
man, who saw in Joseph someone who had received his science directly from God.³ As proof of the objective value and
effect of the science in question, suffice it to say that this divinely inspired knowledge of dream interpretation in the story provided the economic plan of the vast Kingdom of Egypt
for fourteen crucial years. As for reproducibility, Joseph's erst-
while ability to successfully interpret the dreams of his fellow prisoners offers an antecedent to his later interpretation
of Pharaoh's dreams. As regards the verifiability of Joseph's
knowledge, in the Koranic account, before giving his interpretation of his fellow prisoners' dreams, he offered them advance knowledge of events: "He [Joseph] said: The food that
ye are given (daily) shall not come unto you except that I shall
inform you thereof before it cometh unto you. This is of that
which my Lord hath taught me."⁴ It goes without saying that
reproducibility and verifiability take on a much wider meaning in this context than they do in the narrow sense employed
in modern material-empirical science.

To repeat, most important, in the mind of traditional
man, of what can be called science are those bodies of knowledge taught by God directly. Further examples in this context
are useful. In the story of the Queen of Sheba and King Solomon, for instance, one finds insightful points in this regard.
The prelude to the story, in the Koranic account, includes a
reference to an interesting kind of science that God taught
David⁵ and Solomon: "And verily We gave knowledge unto

---

3 "And Joseph said unto them, Do not interpretations belong to God?"
(Gen. 40:8). "Thus we established Joseph in the land that We might teach
him the interpretation of events" (Koran 12:21).

4 Koran 12:37.

5 The Bible provides an example of a God-given science that the Koran
ascribes to King David: "And it came to pass, when the evil spirit from God
was upon Saul, that David took an harp, and played with his hand: so Saul
was refreshed, and was well, and the evil spirit departed from him" (1 Sam.

David and Solomon."[6] The nature of one such science will become clear in the verse that follows: "And he [Solomon] said: O people! We have been taught the language of birds,[7] and have been given of all things."[8] And the subsequent verse notes that with a God-given science and authority, the King ruled with "armies of the jinn,[9] men, and birds." The story, then, has it that, called by King Solomon, the Queen of Sheba is traveling to Jerusalem to pay him a visit at his palace. But before she arrives, the King desires to have her throne moved from her land to his palace in Jerusalem. "I will bring it to thee before thou wouldst rise from thy council,"[10] offers

16:23). The knowledge of "the musical instruments of David the man of God" (Neh. 12:36) were imparted to him by the Lord Himself, for these were "instruments of musick of the Lord, which David the king had made to praise the Lord" (2 Chron. 7:6).

6 Koran 27:15.

7 Knowledge of the language of birds, as strange as it may sound to modern man, is not foreign to the Native American up to this day, for whom the whole of creation is a reflection of God and all creatures members of his greater family. Spirit guides for Native Americans often take the form of animals, including birds. The contemporary Muslim, whose sensibility is conditioned by modernism, and who may find it difficult to understand that "the thunder singeth His [God's] praise" (Koran 13:13), can look to the Native American, for whom thunder beings are not abstract ideas but concrete realities.

8 Koran 27:16.

9 The jinn are conscious beings of the subtle realm, unrecognizable to man by physical means. Like humans, they are responsible for their actions— thus, there are good as well as evil ones—a fact that implies that they are endowed with individual subjectivity as well as free will. Reference was made earlier to nymphs, elves, peris, etc., to name only a few of a very long and varied list of psychic or subtle beings. Speaking of these beings often scandalizes our contemporaries. Even "believers" who cannot deny references to such beings in sacred Scripture would rather not have to deal with the question of their existence. Traditional cosmology, on the other hand, includes immaterial states wherein such supra-sensorial beings find their place.

10 Koran 27:39.

one of the jinn in the service of the King. The King, however, is not impressed. "Said one, who had knowledge of the Book," the subsequent verse notes, "I will bring it to thee before thy gaze returneth unto thee." So he does, the story goes on, and the throne is brought from the land of Sheba to Jerusalem in the twinkling of an eye by the one who possessed a science of a certain "Book" whose author is none but God Himself. The two sciences referred to in the Koranic account of the story, namely King Solomon's knowledge of the language of birds and the King's aide's knowledge of the Book, are certainly beyond the grasp of the modern scientist. Nevertheless, the sciences in question had, in both cases, very concrete—and impressive—effects in the material world.

Beyond the Semitic world, similar examples can equally be found in almost all traditions. According to the ancient traditions of India, for example, Lord Shiva Himself taught the Tamil language to the ancient sage Agastya. Another example would be Osiris, in ancient Egypt, who taught men how to raise corn and vines; Ceres, the goddess of agriculture, had a somewhat similar role for the ancient Romans. And one can go on and on, enumerating scores of examples from diverse traditions.

The example of the agrarian gods of the ancient traditions having taught farming to men is a type of "applied" science. Likewise, the biblical account of the story of Noah presents an "applied" type of science—though of a different kind. The Lord gave Noah instructions about the way the ark was to be constructed and used for accommodating men and animals (Gen. 6:14-16). It is hard to dispute the objective value of this example of "applied" science given the crucial importance it played in saving Noah and his companions during the Flood.

In our discussions on the limitations of reason in an earlier chapter,[11] we noted that the human mind is not limited to

---

11 In chapter 1, "Foundational Questions."

reason alone and benefits from other functions as well—from intuition or insight, for example. We noted the importance of this in refuting the hypothesis of the reconstructibility of the human mind by machines. Careful study of some passages in Scripture provides corroborative references to this truth.

The Bible relates a story in which King Solomon judged between two women who both claimed to be the mother of a child. The King proposed that the baby be cut in two and each woman be given one half. The real mother begged that the child be given to her rival while the other woman had nothing to say. By revealing the feelings of the two women under duress, the King was able to discern their relationship to the child, and thus identified the true mother.[12] The proverbial wisdom of Solomon, as exemplified in this story, was clearly not the result of a mechanistic operation of his rational faculty; rather, it was due to his insight or, in the language of the Bible at the conclusion of the story, it was because "the wisdom of God was in him."

The Koran relates another story wherein King David makes a ruling in a case concerning a field that someone's sheep had overrun, and Solomon offers a better judgment, which, according to the Koran, was made known to him by God: "And We gave Solomon the understanding of the matter."[13] In many situations in life, it seems that judgment or discernment requires more than just discursive thought.

Another important implication of this truth is the inherent insufficiency of the legalistic perspectives that seek solutions to all problems requiring judgment by means of prescribed sets of laws. To discern and judge correctly, one needs the element of insight or inspiration. Plato implies as much when he states that "the law could never accurately embrace

12 1 Kings 3:16-28.
13 Koran 21:79.

what is best and most just for all at the same time,"[14] and "we
ought by every means to imitate the life of the [golden] age of
Cronos . . . in obedience to the immortal element within us."[15]

### Science and the Idea of Progress

The ideology of progress is one that shows a remarkable lack
of imagination:

> If man is intelligent enough to arrive at the "progress"
> which our period embodies—assuming there is any
> reality in such progress—then man must have been
> *a priori* too intelligent to remain for thousands of
> years the dupe of errors as ridiculous as those which
> modern "progressivism" attributes to him; and if he
> is on the contrary stupid enough to have believed in
> them so long, then he must also be too stupid to es-
> cape from them.[16]

Viewed from another angle, this contradiction can be expressed by posing the following question: how can we know
we are progressing if we have no idea of that towards which
we are supposed to be progressing? Needless to say, the same
contradiction undercuts the idea of evolution: how can man,
who is supposedly evolving, suddenly step out of the evolu-
tionary process to make an absolute judgment about this process?

Contrary to the evolutionist hypothesis, all traditions
point to a downward movement of man's spirituality over
time. This is in line with the doctrine of the world cycles,
most explicitly described in Hinduism, according to which

---

14 Plato, *Statesman*, 294a-294b.

15 Plato, *Laws*, 713b-714a.

16 Frithjof Schuon, *Language of the Self: Essays on the Perennial Philosophy*
(Bloomington, IN: World Wisdom, 1999), p. 138.

the four successive stages of a *mahā yuga*, or "great age," correspond to a successive and gradual degeneration of man's spirituality over time. According to this tradition, we are close to
the end of the last stage of the present *mahā yuga*, that is, we
are nearing the culmination of the *kali yuga*, the "dark age," of
our cycle.

For traditional man, the emergence of a new science at a
particular point in time, even if divinely inspired, did not necessarily imply "progress" and a step forward towards further
human advancement—an idea that only gained prominence
during the European Enlightenment, and which still prevails.
In a Koranic verse, on the contrary, God presents the reason
for having taught David how to make a coat of armor: "And
We taught him the making of the suit of armor for your benefit, to protect you from each other's rigor."[17] The fact that at
that particular point in time men had become so violent as to
require protection from one another does not speak so much
of a cultural progress as of a moral decline.

"Writing" is another divine gift to mankind, according
to diverse traditions.[18] However, the advent of scripts to satisfy the need for the preservation of knowledge through writing is a sign of decline in human intellectual capability over
time rather than a sign of advancement. Native Americans
who came into contact with the white man were at times sur-
prised at their need for writing, and wondered if it would not
be enough to hear something to remember it by heart.[19] The
Vedas were put into writing not too long ago. Their learning

---

17 Koran 21:80.

18 Traditionally, the Chinese script is thought to have originated from the
pattern of the footprints of birds, symbolizing angels, and thus implying
angelic inspiration for the origin of the script. In a similar vein, ancient
Egyptian hieroglyphs (literally "sacred writing" in Greek) were reserved for
sacred texts.

19 Which presupposes, of course, that language is not used for vain prattle,
as is so commonly done today, but for communicating things truly worthy
of communication.

is predominantly an oral tradition even today. The fact that Hindu sages felt the need to put them into writing relatively recently, likewise, does not speak in favor of the hypothesis of intellectual progress with time.

Viewed in this light, even the construction of glorious religious monuments and magnificent temples, though divinely ordained and authorized as proper vehicles of God's presence, was a divinely willed adaptation for a humanity that was losing its sense of the metaphysical transparency of phenomena—a humanity which no longer spontaneously saw God everywhere.

In this context, one will note that even the regeneration of spirituality by way of renewed revelations—that is, successive religions—is a divine adaptation of the primordial religion to the conditions of an ever-falling humanity as the world over time moves further and further away from its source. Man in his primordial state was his own prophet. It is his post-edenic condition, however, that necessitates an outward prophet to awaken in him what he knew unaided in his primordial state. Calling this "progress" is like seeing in a frail old man equipped with a staff, eyeglasses, and hearing aids an improvement over his earlier, more integral state.

### Do the Religions Deem All Science Good?

Reference to sacred Scripture, we have said, shows a much wider spectrum for what can be called science than what modern man is willing to admit. In general, all that is called science today would also be considered science in the language of sacred Scripture. But does tradition deem all science good?

On the surface, the answer seems to be in the affirmative, especially for Muslims: "Are those who know equal to those who know not?"[20] asks the Koran; or, "Woe unto him

20 Koran 39:9.

that knoweth not," said the Prophet of Islam, as also: "Seeking knowledge is an obligation upon every Muslim," and "Seek knowledge even unto China." Based on these and similar pronouncements, some have even concluded that the apparent hostility of the Christian Church to the thought of a Galileo contrasts sharply with the character of Islam as a defender of science. Such apologists, however, are not confined to the Muslim world alone, and are found in almost all religious climates today.

Now, to keep things in perspective, it is not only Christ who warns us against too much attention to the here-below at the expense of the hereafter when he says, "For what shall it profit a man, if he shall gain the whole world, and lose his own soul?"[21] The Prophet of Islam also warns against bad science— a science that has worldly benefits in view but is of no use to man's final ends: "God despiseth all those who are knowers of the here-below and are ignorant of the hereafter." Speaking of the ignorant, the Koran says: "They know but the outward in the life of this world but of the hereafter they are heedless."[22] Thus, it seems, there must be a sense of proportion in the extent to which man directs his thought and vital energy to the things of this world. Dedicating one's life to the pursuit of something that has purely worldly gain in mind, even if pursued in the name of science, is of no value in the eyes of God— and thus is of no real and ultimate value to man. Obviously, this does not preclude an activity on his part that is required for his earthly sustenance. Here it is a question of proportion and of right balance. The hypertrophied worldly science of modern man can hardly be presented as a legitimate necessity when considered with man's salvation in view. Material welfare has its place, but it is immeasurably inferior to what man is meant to be: "I created the jinn and humankind only that they

21 Mark 8:36.
22 Koran 30:7.

might serve Me,"[23] God proclaims in the Koran,[24] and Christ commands: "But seek ye first the kingdom of God, and his righteousness; and all these things shall be added unto you."[25] We conclude, then, that the sciences that have worldly gain in view, even if they are not bad in themselves, may become so, by traditional measures, when the proper balance is lost. Traditional worlds were aware of this, and care was normally taken not to over-accentuate the worldly at the expense of the spiritual. For instance, Rumi says:

> He [the worldly man] struggles more than the beasts,
> (for) he performs subtle arts in the world.
> The cunning and imposture which he knows how to
> spin—
> that (cunning) no other animal produces.
> To weave gold-embroidered garments,
> to win pearls from the depths of the sea,
> The fine artifices of geometry
> or astronomy, and the science of medicine, and philoso-
> phy—
> Which are connected only with this world
> and have no way (of mounting) up to the Seventh Heav-
> en—
> All this is the science of building the (worldly) stable
> which is the pillar [basis] of the existence of (persons
> like) the ox and the camel.
> For the sake of preserving the animal [in them] for a few
> days,

---

23 To which the Prophet of Islam added, by way of commentary: "meaning, that they might know Me." Man serves God best by use of the best of his faculties, namely his intelligence. Therefore, to "serve" God best means to "know" Him, which by way of consequence means to "will" Him and to "love" Him.

24 Koran 51:56.

25 Matt. 6:33.

these crazy fools have given to those (arts and sciences) the name of "truths."[26]

We noted that, according to traditional measures, the pursuit of a science that has worldly gain as its sole purpose is of no value in the eyes of God, and ultimately for man as well. Worldly gain, however, is not the only objective that could render a science worthless: that which is useless, as regards man's reason for being and his final ends, will likewise make a science blameworthy when it becomes its subject matter. Spending a lifetime on researching myriads of things of this world with no real worldly or spiritual benefit is so characteristic of the scientific spirit of our time that the question of the worth of such vain endeavors hardly arises. Scientific curiosity clashes head-on with what tradition teaches: "O God! I take refuge in Thee from a science that hath no benefit," said the Prophet of Islam on one occasion, and "Part of the perfection of one's Islam is his leaving that which does not concern him," he said on another. Man's life is too precious, and too short, to be spent on useless things. In fact, what is useless may become harmful: "He that is not with me is against me,"[27] said Christ. And, in speaking of the people of a good state, Lao Tzu tells us that "even though the next country is so close that people can hear its roosters crowing and its dogs barking, they are content to die of old age without ever having gone to see it,"[28] thus enjoining us to beware of unhealthy curiosity.

We have seen that, according to tradition, a science may be blameworthy even if its subject matter is not evil in itself. But it may also happen that the content of a science is evil

---

26 *Mathnawī*, Book IV, 1527-1533.

27 Matt. 12:30.

28 *Tao Te Ching*, chapter 80 (Stephen Mitchell translation). It goes without saying that this quotation is not meant to be taken literally outside its particular context; nevertheless, the spirit of Lao Tzu's words is universal and thus applicable to all circumstances.

in itself. The Prophet of Islam said, "God, His angels, the dwellers of the heavens and the earth, and even the ant in its hole and the fish in [the depths of] the sea give blessings upon those who teach people knowledge." And yet it is hard to imagine that the fish in the depths of the sea where nuclear tests are being conducted would bless those who teach how to conduct such tests! The teacher of some sciences, according to sacred Scripture, may be the devil[29] himself or other evil spirits. Sorcery, necromancy, fortune-telling, and divination—occult practices that have always infested humanity, especially during periods of decadence[30]—are all sciences well attested to in sacred Scripture,[31] and are declared forbidden in the most severe terms by all authentic religions;[32] far from being mere superstitions, they too may have concrete results.[33]

29 As we shall see in the next section, the infinitude of the Principle necessitates a potential for its own negation, whence the possibility of the tendency towards nothingness. The devil is the personification of the cosmic tendency towards nothingness when it comes into contact with man.

30 There is no shortage of these kinds of sciences and scientists in our time, albeit sometimes with new names: spiritism, after-death communication, astral projection, out-of-body experiences, cosmic consciousness, altered states of consciousness, psychic reading, etc.; common in parapsychology and paranormal circles, they are merely new terms for old concepts.

31 For example, "Regard not them that have familiar spirits, neither seek after wizards, to be defiled by them: I am the Lord your God" (Lev. 19:31). "There shall not be found among you any one that maketh his son or his daughter to pass through the fire, or that useth divination, or an observer of times, or an enchanter, or a witch. Or a charmer, or a consulter with familiar spirits, or a wizard, or a necromancer. For all that do these things are an abomination unto the Lord: and because of these abominations the Lord thy God doth drive them out from before thee" (Deut. 18:10-12).

32 For example, "A man also or woman that hath a familiar spirit, or that is a wizard, shall surely be put to death: they shall stone them with stones: their blood shall be upon them" (Lev. 20:27).

33 "Then Pharaoh also called the wise men and the sorcerers: now the magicians of Egypt, they also did in like manner with their enchantments" (Exod. 7:11). "And the magicians did so with their enchantments, and brought up frogs upon the land of Egypt" (Exod. 8:7).

That these were sciences taught to mankind by evil spirits—
"by God's leave," of course—is reflected well in the following
passages from the Koran: King Solomon had "spirits, every
kind of builder and diver"[34] in his service; still "Solomon disbelieved not; but the [evil] spirits disbelieved, teaching mankind magic and that which was revealed to the two angels in
Babylon, Harut and Marut. Nor did they teach it to anyone
till they had said: 'we are only a temptation, therefore disbe-
lieve not.' And from these two, people learn that by which
they cause division between man and wife; but they injure
thereby no one save by God's leave. And they learn that which
harmeth them and profiteth them not. And surely they do
know that the buyers [of magic] will have no [happy] portion
in the hereafter; and surely wretched is that for which they
sell their souls, if they but knew."[35] Many things are science
then, according to tradition, but not all that is science is good.
As the wise men of old would say, "the nobility of a science is
of the same measure as the nobility of its subject matter."

Another example of a science that is forbidden to man
is found in the Koran in a dramatic scene where the devil,
after having been cursed and cast out of the Garden of Eden,
threatens God by saying that he will take of God's "servants
an appointed portion";[36] he then goes on to describe his strat-
egy for doing so: "And surely I will lead them astray, and surely I will arouse desires in them, and surely I will command
them and they will slit the ears of cattle, and surely I will command them and they will change God's creation."[37] One wonders what this "changing of God's creation," so strongly condemned in this passage as a major satanic stratagem, might be
in our time, if not genetic engineering, which aims precisely at
"changing God's creation."[38]

34 Koran 38:37.

35 Koran 2:102.

36 Koran 4:118.

37 Koran 4:119.

38 Futuristic science is not likely to stop with cloning animals, and one

### Epistemological Doctrines of the Religions

Implicit in the messages of all the religions is man's innate ability to recognize their truth: if we did not inherently possess the truths to which the religions call us, how could we possibly accept them? The degree to which different religions make their epistemologies explicit varies though, and in each case it depends on the structure of the religion in question.

A full study of the epistemological doctrines of all the religions is beyond the scope of this book. Instead, an example of these epistemological doctrines, taken from the Koran, is presented here in some detail, and is then followed by a brief presentation of epistemological references in some other traditions. The sacred book of Islam is chosen here for an extensive study because we find in it explicit references to epistemology—a doctrine that no other authentic religion lacks, of course, however implicit it may be in its way of expressing it.

It should also be remembered that the language of sacred Scripture is symbolic. While the doctrines that the wise men of every tradition expound derive directly or indirectly from the Scripture of that tradition, Scripture itself uses a very synthetic and concise language full of symbolism, which puts it on a far higher level than that of the analytical expositions of sapiential doctrines. Moreover, passages from sacred Scripture are not philosophical narratives, each centered on a single idea; as such, they are rich in digressionary remarks, worthy of note in their own right, which we should not pass over in silence.

Koranic epistemology may be inferred from a striking passage about creation in which the spectacle of man's creation is presented in an extraordinary scene. Here, God, His angels, and the spirits of every kind, even the devil, are present. The passage begins with God's proclamation of His intention to create man: "And when thy Lord said unto the an-

---

cannot but be concerned about the inevitable emotional and psychological crises of a cloned human being, let alone the question of the nature of the spirit of a cloned human.

gels:'Lo! I am about to place a vicar on earth.'"[39] God's vicar, or representative, on earth is what, in God's own words, defines man here. To be able to represent well him whom he represents, the representative will have to resemble him as closely as possible. That is why, in the language of the Bible, "God created man in his own image."[40] In both the biblical and Koranic accounts of the successive stages of creation, man appears last on earth. He becomes the keystone that completes the construction of the edifice of creation, and is its reason for being. Man's deiform nature accords him certain prerogatives: an intelligence that is total, hence capable of objectivity, which enables him to conceive of the Absolute; a free will that enables him to choose what may be contrary to his immediate individual interest; and a soul (or sentiment) in conformity with his objective intelligence and free will, that is, capable of disinterestedness and generosity.[41] Needless to say, all creation is a reflection of God: "I was a hidden treasure and I wished to be known; therefore, I created the world"; according to this Divine utterance from the Islamic tradition, the whole world is a manifestation of God.

Every being, animate or inanimate, is prefigured in and emanates from a Divine Name, that is to say, it is a reflection of an aspect of God on a particular plane of existence.[42] Man, however, manifests the Divine Name itself according

39 Koran 2:30.

40 Gen. 1:27.

41 For a thorough treatment of this subject, one is referred to Frithjof Schuon, "Prerogatives of the Human State," *The Play of Masks* (Bloomington, IN: World Wisdom, 1992), pp. 1-16.

42 The Divine Intellect permeates all creation down to the domain of inanimate objects. Every being partakes of the Intellect: the intelligence of animals manifests itself through their instincts; as for plants, the turning towards the sun of the sunflower, for example, is a manifestation of its "intelligence"; even in the domain of minerals, the form of a diamond, for instance, is an expression of its "intelligence." Reference is made in this context to Frithjof Schuon, *The Transcendent Unity of Religions* (Wheaton, IL: Quest Books, 1984), pp. 56-57.

to its aspect of totality.[43] His will, manifesting God's, must therefore be free. In this, the angels—who are aspects of the Divine Intellect,[44] and, as such, see in every possibility its existential unfolding—perceive the peril: "Wilt thou place therein [on earth] one who will cause corruption therein and shed blood?"[45] Because man is created according to the Divine Name in its aspect of totality, he stands at the center of the plane of existence that is his home, namely, the terrestrial world. Precisely because he is situated on the vertical line that passes through his plane of existence, man can either ascend upwards to Heaven or fall downwards. This is what the angels foresee in the possibility of a being in possession of freedom placed, not in God's proximity in Heaven as are the Archangels and the blessed, but on earth, that is, in remoteness from the Divine. The freedom thus accorded to man gives him his deiformity, his quality of being God's vicar on earth, but at the same time gives him the ability to choose what may be contrary to this vocation and, in the final analysis, contrary to his intelligence. An animal's will is instinctive; therefore, it cannot sin. Man, on the other hand, can sin—and can do so dramatically[46]—whence the angels' apprehension. Once God's vicar on earth chooses to be unfaithful to his very substance, the whole terrestrial world is corrupted. *Corruptio optimi pessima*—"the corruption of the best is corruption at its worst."

43 According to the Divine Names *Allāh* or *Al-Rahmān*, in the Koran.

44 Angels are the celestial manifestations of the Divine Names just as earthly beings are their terrestrial manifestations. And just as man is the earthly manifestation of the Divine Name in its aspect of totality, the Spirit of God—the Divine Word or the Divine Intellect along with its heavenly persons, that is, the Archangels—is its heavenly manifestation. Peripheral angels are aspects of the Divine Intellect whereas the Spirit of God is the Divine Intellect in its totality. Likewise, the intelligence of non-central beings—on earth: animals, plants, and minerals—is partial whereas that of the central being—on earth: man—is total.

45 Continuation of the same verse, that is, Koran 2:30.

46 As Schuon notes in *The Play of Masks*, p. 20.

The language of Scripture is symbolic, and symbolism by its nature has many levels. Accordingly, one should not take the Koranic account of creation presented here literally, as if it referred to a historical event. Anthropomorphic[47] imagery offers every man an understanding of its truth according to his level of comprehension. Scripture does not solely address the man of analytical and contemplative mind, but offers the fundamental truths in ways accessible to every man. Yet, the most profound doctrines also find their prefiguration in Scripture. Moreover, by virtue of being the Word of God, revealed Scripture vehicles God's presence, whence its sacramental character.

The potential of man's downward fall and perversion because of his free will, foreseen by the angels in the cited Koranic passage, is related to the problem of evil. Much has been made of this, leading some to go so far as denying God altogether: for if God is truly omnipotent and good, why would He not eradicate evil from the world? If He cannot eradicate evil, is He truly omnipotent? And if He will not, is He truly good? But there is a metaphysical explanation for this conundrum,[48] an outline of which is as follows.

By virtue of being absolute, God is infinite. Any limitation ascribed to Him takes away from His absoluteness; therefore, absoluteness implies infinitude. Likewise, any determination ascribed to Him takes away from His limitlessness; therefore, infinitude implies absoluteness. The Absolute and the Infinite are in fact identical; and if there is a duality, it is then solely in our mind. Infinitude, in turn, implies All-Possibility, that is, the principle of all that is possible;

47 Anthropomorphic language is valid and plausible precisely because man is made in the image of God.

48 Which is best given by Schuon in several of his books, notably in Frithjof Schuon, "The Question of Theodicies," *The Eye of the Heart*, pp. 31-45, and in Frithjof Schuon, "Dimensions of Omnipotence," *Survey of Metaphysics and Esoterism* (Bloomington, IN: World Wisdom, 1986), pp. 65-76.

it also implies that these possibilities manifest themselves on the plane of existence to the extent this can be actualized. But, to manifest these possibilities at a lower plane, infinitude requires a potential for its own negation, which it has in the possibility of nothingness; a nothingness that is lent an illusory existence through Being, for pure nothingness can only exist as a kind of tendency, that is, as a privation of Being, or of the Good. Evil is only a form of this privation of the Good, that is, as it enters the formal state. A world without evil, illness, old age, and death is a possibility, as is a world with them. The former exists—it is Paradise—but so too must the latter, whence our terrestrial world.[49] To wish that there be no evil in our world is to wish that there be no such world. Evil is a necessary constituent of our world. Its existence is the trifle of a price to be paid in view of the greater good it brings with it. That is why God responds to the angels' apprehension, saying "Surely, I know that which ye know not."[50]

"And He [God] taught Adam all the names,"[51] the Koranic account continues. The "names" are the principial realities that lie at the foundation of every being, in its potentiality or in its existentiated form. Unlike peripheral beings whose allotment of the "names" is partial, man received "all the names," whence his aspect of totality, which confers on him the quality of being God's vicar. This "teaching" of the "names," furthermore, takes place before man's creation on earth, since it is only after the passing of these "events" that God says, "O

---

49 "Evils, Theodorus, can never be done away with, for the good must always have its contrary; nor have they any place in the divine world, but they must needs haunt this region of our mortal nature" (Plato, *Theaetetus*, 176a).

50 Continuation of the same verse, that is, Koran 2:30.

51 Koran 2:31.

Adam! Dwell thou and thy wife in the Garden[52] [of Eden]."[53] The story is, of course, a mythological account; in metaphysical terms, it means that fundamental truths are imprinted in man's spirit, and, given that this precedes man's creation in time, it signifies that man's spirit has priority over his terrestrial existence.

Man was "taught" the knowledge of all fundamental truths[54] prior to his creation: this is the very thesis of Plato's *anamnesis*—though expressed differently. In Rumi's words:

> We all have been parts of Adam,
> we have heard those melodies in Paradise.
> Although the water and clay [of our bodies] have caused
>     some doubt to fall upon us,
> we (still) remember something of those [melodies].[55]

Man can know things because their principial realities are imprinted in his very spirit. The angels, in the same passage, expressed this very idea when they said "of knowledge

---

52 Eden was the terrestrial garden in which man was placed after his creation. Man erred in this garden whereas there is no possibility of error and sin in God's bosom, that is, in Paradise proper. The existence in this garden of the serpent—Satan—is another indication of the existence of evil there. Evil has no place in the proximity of the Divine, that is, in Heaven.

53 Koran 2:35.

54 Fundamental truths are expressions of eternal ideas in the human spirit. "Fundamental," because all secondary truths are merely their repercussions. As an example, on a very particular plane of the mind, to possess Euclid's five postulates is to possess the whole of Euclidean geometry: in this example, the five postulates are fundamental while all the theorems of Euclidean geometry are secondary. The "enlightened" man does not necessarily know all the contingent facts of the terrestrial world, as the philosopher Russell would surmise, but his knowledge of the fundamental truths is actual in him, and, as such, he sees in every phenomenon its principial reality.

55 *Mathnawī*, Book IV, 744-745.

we have none, save what Thou hast taught us,"⁵⁶ after God "showed them (all the beings) to the angels, and asked them: 'Inform Me of the names of these, if ye are truthful.'"⁵⁷ The peripheral angels are partial manifestations of God's Intellect, hence they could not do what man's spirit, total in its mani-festation of God's Intellect, was able to do: "He [God] said: 'O Adam! Inform them of their [all the beings'] names,' and when he had informed them of their names, He said: 'Did I not tell you that I know the secret of the heavens and the earth?'"⁵⁸ In these verses, "all the beings" refer to the totality of all elements of existence, which man was able to recognize— that is, was able to know—because he had been "taught" their names, meaning that he possessed their principial realities in his spirit.⁵⁹ And given man's totality, God commands the pe-ripheral angels to prostrate themselves before him: "We said unto the angels: Prostrate yourselves before Adam, and they fell prostrate, all save Iblis [the devil].⁶⁰ He refused and was proud, and became a disbeliever."⁶¹ Notice that it was the pe-ripheral angels and spirits who prostrated themselves before Adam and not the Spirit of God.⁶²

In another Koranic account of the "events" before cre-ation, explicit reference is made to the fact that all men, prior to being born on earth, have borne witness to the truth of

56 Koran 2:32.

57 Koran 2:31.

58 Koran 2:33.

59 In the Perfect Man the knowledge of realities is actual; in fallen man it is merely potential or virtual.

60 Satan, or Iblis, was a being of the subtle realm, and not an angel: "Iblis was of the jinn," the Koran specifies in another passage on creation (Koran 18:50).

61 Koran 2:34.

62 The Koran makes a distinction between *Al-Rūh*, the "Spirit of God," and the peripheral angelic beings in a number of places, for example in 17:85, 78:38, and 97:4.

God: "And when thy Lord brought forth from the children of Adam, from their loins, their seed, and made them testify of themselves, (saying):'Am I not your Lord?' They said:'Yea! verily. We testify.'"[63] In this account, "seed" indicates potentiality, that is, a state of man antecedent to his earthly incarnation. That man has seen God before coming into this world is a symbolical expression of the fact that he carries eternal realities in his spirit. This makes him accountable before God, for he who carries within himself the innate knowledge of the Divine never ceases to be responsible for his vocation, which is also his reason for being, whether he acknowledges it in his lifetime or not: "Lest ye should say on the Day of Judgment: 'Lo! of this we were unaware.'"[64] "Or lest ye should say: 'Our fathers took before us false gods and we were (their) descendants after them. Wilt Thou destroy us on account of that which those who follow falsehood did?'"[65] In other words, we know the eternal truths deep down and need only to remember them in this world. This explains why the Koran does not speak of "knowing" God, but on the contrary always speaks of "remembering" Him. And, that is also why in the Koran man's fallen state is called a state of "forgetting." The phenomena of this world and his own soul are there to reawaken in him what he knew in his spirit: "We shall show them Our signs on the horizons and within themselves until it becomes manifest unto them that this is the Truth."[66]

That for the wise men of old, man's faculties of knowledge were not limited to reason alone can very well be seen in the following words of Ali, the prominent companion of the Prophet of Islam: "At all times, periods after periods, even during the periods of [spiritual] lassitude, there are people to

---

63 Koran 7:172.

64 Continuation of the same verse, that is, Koran 7:172.

65 Koran 7:173.

66 Koran 41:53.

whom God whispers through their minds and speaks through their innermost intellect."[67] The human mind, whose organ is the brain, is different from the intellect, whose seat is traditionally considered to be the "heart," that is, man's spiritual center. Whereas our mind can only receive a reflection of the truth, our heart, the seat of the intellect, is capable of direct communion with the truth, whence the direct vision of the "eye of the heart." Ali's words here are carefully chosen, and refer to two different, hierarchical levels of comprehension, one cerebral, and another cardiac: whispering through the mind—hence referring to a cerebral or mental understanding—can only give an indication of what is openly spoken through the intellect—hence referring to a cardiac or intellectual intuition.

The parable of the elephant in the dark[68] presents this same hierarchical relationship by way of an analogy: rational thought is likened to the touching in the dark, one by one, of various parts of the elephant in order to form a picture of what it might be in terms of what one knows; this in contrast to intellection, which amounts to a direct and total vision of the elephant in broad daylight. The intellect is what endows man with objectivity. Man's intelligence is objective because, deep in his spirit, every man has access to the intellect, which is thus universal. It is because of this that man can recognize the truth independently of his own subjectivity. The very fact that men communicate with one another, and understand each other, is indicative of the existence of common, universal truths, to which all men have access. The very fact that some

---

67 *Nahj al-Balāghah* ("Peak of Eloquence"), Sermon 222.

68 Or the parable of the blind men and the elephant, well-known, in various forms, in oriental traditions. In one such account, an elephant is exhibited in a dark room. A group of men touch and feel various parts of the elephant in the dark, and, depending on where they touch it, believe the elephant to be in the like shape: a water spout, a fan, a pillar, and a throne—by those who touched the trunk, the ears, the legs, and the back of the elephant respectively.

present theories, in the expectation that others will accept them, is indicative of their tacit assumption of the universal intellect, even if their theories be the very antithesis of such an assumption.

The intellect, "the immortal element within us"[69] according to Plato, is that "something in the soul that is uncreated and uncreatable," in the words of Meister Eckhart. "The Universal Intellect was the first thing God made," said the Prophet of Islam. Reason is a reflection of the intellect in the human mind and, precisely because of this very fact, is capable of pointing to its own limitation with respect to its inexhaustible source.

What was blameworthy in the eyes of the wise men of old was not the rational faculty itself—which was considered not only good but also necessary on its own plane—it was, rather, the exclusivist claim of rationalism that would make truth a province of reason alone.[70] So precious is logic,[71] on its own level, that even God, when He speaks to man in Scripture, is logical.[72] Scripture, and along with it the written wisdom of the sages throughout the ages, uses logic, not as the exhaustive criterion of truth, but as a means of awakening in man what he bears in his immortal kernel.

"Behold, the kingdom of God is within you,"[73] Christ teaches us, which prefigures all the other epistemological references in Christianity, for example, where he says, "For it is

---

69 Plato, *Laws*, 714a.

70 It is this that Rumi had in mind when he said, "The leg of the rationalist is of wood / a wooden leg is very infirm" (*Mathnawī*, Book I, 2184). Rumi's *Mathnawī* itself is presented with the aid of reason.

71 According to St. John of the Cross, right reason is the temple of God (*The Collected Works of St. John of the Cross*, third edition, translated by Kieran Kavanaugh and Otilio Rodriguez [Washington, DC: ICS Publications, 2017], p. 139).

72 The apparent contradictions in sacred Scripture are in reality ellipses that call for explanations, which the traditional commentaries provide.

73 Luke 17:21.

not ye that speak, but the Spirit of your Father which speaketh in you."[74] And St. Gregory Palamas says of the totality of man's innate knowledge: "Man, this greater world in little compass, is an epitome of all that exists in a unity and is the crown of the divine works."[75] Likewise, Jacob Boehme who says, "Man alone contains within himself as many species as exist on earth."[76] Meister Eckhart too reflects on the fact that all creaturely knowledge is contained in man's spirit, when he says: "Well I ween that if I knew myself as intimately as I ought, I should have perfect knowledge of all creatures."[77] Clearly, then, man is God's vicar on earth because he reflects God in His totality: "Angels and all other Creatures have their destined Ideas in the *Divine Mind*. But God himself in his own essential Image, in the Person of the Son, the *Idea of the Ideas*, is the *Idea of Man*."[78]

Turning to Judaism, the account of man's creation and his knowledge of the "names" in the Bible prefigures much of what the Koran relates, conveying essentially the same doctrine: "God created man in his own image,"[79] and, therefore, he has the knowledge of the "names": "out of the ground the Lord God formed every beast of the field, and every fowl of the air; and brought them unto Adam to see what he would call them: and whatsoever Adam called every living creature, that was the name thereof."[80]

---

74 Matt. 10:20.

75 Quoted in Titus Burckhardt, *Introduction to Sufi Doctrine* (Bloomington, IN: World Wisdom, 2008), p. 65.

76 Jacob Boehme, *Mysterium Magnum*, translated by Nicolas Berdiaeff (Paris, France: Aubier, 1945), XX.34.

77 *Meister Eckhart*, by Franz Pfeiffer, translated by Clare de Brereton Evans (London, UK: John M. Watkins, 1924), I.324.

78 Vivian de Sola Pinto, *Peter Sterry: Platonist and Puritan* (London, UK: Cambridge University Press, 1934), p. 97.

79 Gen. 1:27.

80 Gen. 2:19.

One finds the same principles in Hermetic doctrines, which represent the sacerdotal wisdom of Egyptian antiquity: "If you possess true knowledge, O Soul, you will understand that you are akin to your Creator";[81] thus speaks Hermes Trismegistus of man's deiformity, which enables him to have full knowledge of everything. He says, moreover, "You ought, O Soul, to get sure knowledge of your own being, and of its forms and aspects. Do not think that any one of the things of which you must seek to get knowledge is outside of you; no, all things that you ought to get knowledge of are in your possession and within you. Beware then of being led into error by seeking [elsewhere] the things which are in your possession."[82] Science, then, is nothing but recollection: "*Scientia* is to be got by the mind being called back to itself and gathered together into itself."[83]

The oriental doctrines of knowledge are likewise based on this same quality of man as being God's direct, hence total, reflection. A Hindu text expresses it thus: "Human birth ... reflects my image."[84] And creation is nothing but an existentiation of the possibilities: "Creation is only the projection into form of that which already is."[85]

Man has all the fundamental truths within himself and should not look elsewhere. In Lao Tzu's words, "One who knows others is clever, but one who knows himself is enlightened."[86]

---

81 *Hermetica*, edited and translated from the Greek and the Latin by Walter Scott (Oxford, UK: Clarendon Press, 1924-1936), *De Castogatione Animae*, 9.4.

82 Ibid., 12.5.

83 Ibid., 5.6.

84 *Srīmad Bhāgavatam*, XI.19 (Swami Prabhavananda translation).

85 Ibid., III.2.

86 *Tao Te Ching*, chapter 33 (D.T. Suzuki and Paul Carus translation).

One need not go further. All the great traditions spring forth from the same divine source and bear witness to the same eternal truth.

## Traditional References to the Multiple States of Being

The traditional doctrines of all the great religions account for varied and graded degrees of reality. Forms of presentation may vary, however, from one traditional perspective to another, and, depending on the scale of gradation desired, different accounts may be given.

The five Divine Presences, presented earlier as the basis of our illustration of the degrees of reality, are referred to in Sufi terminology, for example, as *Hāhūt* (the Divine-in-Itself, the Infinite Self), *Lāhūt* (the Divine Realm), *jabarūt* (the realm of power), *malakūt* (the realm of royalty), and *nāsūt* (the human realm) or *mulk* (kingdom).

"I was a hidden treasure and I wished to be known; therefore, I created the world." According to this Divine utterance coming from the world of Islam, creation is a manifestation of God. "That which is below is like that which is above and that which is above is like that which is below to do the miracles of one only thing," we read in the *Emerald Tablet*, which is the essence of the Hermetic teachings. Each successive degree of Manifestation projects the realities of the preceding degree onto a realm further removed from the divine source: "God the Almighty created His *mulk* [gross state] on the pattern of His *malakūt* [subtle state], and founded His *malakūt* [subtle state] on the pattern of His *jabarūt* [supra-formal state] to lead from His *mulk* [gross state] to His *malakūt* [subtle state], and from His *malakūt* [subtle state] to His *jabarūt* [supra-formal state]."[87]

According to a tradition of the Prophet of Islam: "The earth [material world] and the celestial spheres [sensorial

87 A saying of Jafar al-Sadiq, a prominent eight-century saint of the world of Islam.

world] and all that there is therein are but a ring thrown in the desert compared to God's footstool [subtle world], and God's footstool is but a ring thrown in the desert compared to His Throne [supra-formal Manifestation], and His Throne is but a ring thrown in the desert . . . and all that in God's hold is like a seed in a man's hand, or less." The degrees of Manifestation are here referred to by the words "earth," "celestial spheres," "footstool," and "Throne." These words, Koranic in origin, refer, respectively, to the material world, the sensorial world (the two worlds together forming the gross state), the subtle world, and the supra-formal state.

"The Clement is established on the Throne"[88] and "His footstool encompasseth the celestial spheres and the earth."[89] The anthropomorphic-mythological imagery here is implicitly that of a King [God] seated on His Throne and resting His feet on a footstool, and is likewise found in the Bible: "Thus saith the Lord, The heaven is my throne, and the earth is my footstool."[90] Opposition begins to appear only where duality appears, that is, at the level of the footstool (on which the *two* legs rest). Beyond that, that is, in Heaven, there is no opposition, and no evil.

"God is the light of the heavens and the earth. The symbol of His light is as a niche wherein is a lamp and the lamp, in a glass. The glass is as it were a shining star. (The lamp is) kindled from a blessed tree, an olive neither of the east nor of the west, whose oil is well-nigh luminous though fire touched it not. Light upon light. God guideth unto His light whom He will. And God speaketh to mankind in symbols, for God is the Knower of all things."[91] The verse of light, as this is called, offers, in a symbolic language, a full ontology. God is the Principle. His light pervades the whole of Mani-

---

88 Koran 20:5.

89 Koran 2:255.

90 Isa. 66:1. "The earth" here stands for the gross state in its entirety.

91 Koran 24:35.

festation, which is then described as the image of an oil lamp in a glass enclosure housed in a niche;[92] the image, furthermore, is evocative of the hierarchical degrees of existence in the Macrocosm—the supra-formal, subtle, and gross states, respectively—as also in the microcosm—spirit (or intellect), soul, and body. Man's intellect receives its light from the Universal Intellect—the blessed olive tree that is "neither of the east nor of the west," which is to say that it is formless. But man's intellect contains the divine realities within itself already—"it is well-nigh luminous" in itself. Revelation thus awakens the inner truths that are contained in man's intellect—whence "light upon light"—by furnishing symbols that open to the truths of higher realities, and that is why "God speaketh to mankind in symbols."

The doctrine of the multiple states of being is found in various degrees of explicitness in all other great traditions.[93] The Sanskrit term *Triloka*, literally "three worlds," refers to the three degrees of being in Manifestation; it thus presents a full account of the Hindu cosmology. *Triloka*, when taken together with the Principle, *Brahma*, represents a full account of metaphysics in Hinduism. Within the Divine Order, the Essence, *Nirguna Brahma*, literally "unqualified *Brahma*," is distinguished from its first self-determination, *Saguna Brahma*, literally "qualified *Brahma*."

We saw that the lower two degrees in the doctrine of the five Divine Presences can be taken to represent one degree, the formal or natural world, representing only a part of Mani-

92 The original Arabic word (*mishkāh*) can refer to a prayer niche or a perforated metal enclosure; both are possible translations.

93 Although this doctrine may not be explicitly expressed in a given tradition, it is implicitly assumed in its perspective. Witness the Native Americans whose outlook comprises all these degrees while not having been expressed as a written doctrine. In the sun dance, for example, the central tree of the sacred lodge has three rings painted on it, representing the gross, subtle, and divine realms.

festation, thus providing a description of total Reality by four degrees: the Principle, the prefiguration of Manifestation in the Principle, the reflection of the Principle within Manifestation, and Manifestation proper. The following accounts relate to this fourfold image of reality.

The Far-Eastern symbol of the *Yin-Yang*[94] may be interpreted in such a way as to provide a doctrine of reality in visual form, in keeping with the predominantly visual nature of the Far-Eastern spirit. According to one interpretation,[95] the two poles in this symbol are envisaged hierarchically, the white field representing the Principle, and the black field Manifestation. The black dot in the white field is the prefiguration of Manifestation in the Principle, that is to say the first self-determination of the Essence, while the white dot in the black field is the reflection of the Principle in Manifestation, namely the Universal Intellect. All metaphysics can be inferred from this symbol: If Manifestation did not have its prefiguration in the Principle, it could not come into existence. Likewise, if the Principle were not present in Manifestation, the world would be naught. The two fields are inseparable, which means that to say God is to say Creation, and vice versa. The meanings in this symbolism are inexhaustible, which explains why this image can be the object of years of contemplation for the Taoist contemplative.

94 The *Yin-Yang* is the principle of compensatory reciprocity. The two poles of *Yin* and *Yang* attract and complement one another. Each has an element of the other. In the horizontal relationship, neither pole is superior to the other, and the right balance between the two constitutes harmony. *Yin* is the passive or feminine pole and may represent, for example, such qualities as cold, dark, and black, and may denote such things as water, north, moon, and earth, in contrast to *Yang*, which is the active or masculine pole and may represent, for example, such qualities as warm, light, and white, and may denote such things as fire, south, sun, and heaven.

95 In this connection, reference is made to Frithjof Schuon, *Form and Substance in the Religions*, p. 208.

A similar symbolism may be found in the introductory words of the Gospel of St. John, which mirrors that of the *Yin-Yang* in verbal form: "In the beginning was the Word, and the Word was with God, and the Word was God." Temporal priority here symbolizes principial priority: "In the beginning" (*in principio*) refers to the Divine Principle. The Word that "was with God" and "was God" is the self-determination of the Principle, that is, the Uncreated Logos, which is at once the prefiguration of Creation in the Principle and the ontological cause of Creation, which is why it is said that "all things were made by him; and without him was not any thing made that was made." "And the light shineth in darkness; and the darkness comprehended it not": it is in the nature of the Principle to radiate into Creation and it is in the nature of certain elements in Creation to resist it; otherwise, the world would not be the world.[96] The reflection of the Principle in Creation is the Word that "was made flesh." This is the created Logos, which was "created" before creation, the word "create" here having a higher and transposed meaning.[97]

Parallel to the symbolism of the *Yin-Yang* interpreted as a visual representation of metaphysical principles, the four successive degrees of reality may be inferred from the four words of the first testimony of faith in Islam, "*lā ilāha illa'Llāh*," that is, "there is no god but God." The first two words (*lā ilāha*), literally "no god," constitute a negation, and thus represent Manifestation. The last two words (*illa'Llāh*), literally "but God," constitute an affirmation, and thus represent the Prin-

---

96 See Frithjof Schuon, *Roots of the Human Condition*, p. 79.

97 This priority of the created Logos over Creation can also be seen in this Divine utterance from the world of Islam (addressed to the Prophet): "Had it not been for Thee, O Ahmad, I would not have created the Universe." "Ahmad" is the name of the Prophet of Islam in Heaven, that is, it refers to the Logos. Witness also this saying of the Prophet: "I was a prophet while Adam was between water and clay; and I was a prophet when there was no Adam, no water, and no clay."

ciple. Within Manifestation, *ilāha*, that is "god," is the reflection of the Principle, and in the Principle, *illā*, that is "but," is the prefiguration of Manifestation.[98] The first testimony of faith in Islam is thus the supreme discernment between the Real and the illusory, much like the Vedantic formula "*Brahma is real, the world is illusory*," which, followed by "the soul is none other than *Brahma*," establishes the same reciprocity between the Principle and Manifestation.

Likewise, by one possible interpretation that parallels the above, one can see the same doctrine in the initial four verses of chapter 55, "The Clement" (*Al-Rahmān*), in the Koran: "(1) The Clement (2) made known the Koran, (3) created man, (4) taught him speech." "The Clement," or "the infinitely Good" is the Supreme Principle; "made known the Koran" represents the first self-determination in the Principle since this "making known" is prior to creation, the Koran being the Divine Word; "created man," that is, He projected Himself *ab extra*, whence Manifestation; "taught him speech": "speech" is the prerogative of man and is indicative of the ray of the Universal Intellect in him, thus representing the reflection of the Absolute in man, the summary of all creation.

In this order of ideas, it may be helpful to look at some references in the words of the saints. It is of the Principle that Ali speaks, in apophatic terms, in the first sermon of the *Peak of Eloquence*, when he makes reference to "He for whose reality no limit may be conceived, and whom no attribute can describe," while the knowledge of Him is perfected only when "all attributes are denied Him" for "whoever describes Him, thereby limits Him." "He is, but not by having come into being. . . . He is with everything, but not in likeness. He is separate from everything, but not by remoteness." In the latter, we also find the notions of transcendence and immanence: God transcends all, yet is immanent in all. Likewise, in the words

98 See Frithjof Schuon, *Sufism: Veil and Quintessence* (Bloomington, IN: World Wisdom, 1981), pp. 133-134.

of the thirteenth-century Sufi saint Ibn al-Farid, who here speaks for the Essence: "One who praiseth Me succeedeth in [his] praise of Me by praising My Attributes through Me; otherwise, to praise Me by Attributes is to demean Me."

The "unqualified" Essence is one, whereas "qualification" engenders indefinite veils. "I am as My servant thinks I am," according to a Divine utterance from the world of Islam. According to a tradition of the Prophet of Islam, on the Day of Judgment God will appear to the believers "in a likeness closest [to non-qualification]." When God tells them, "I am your Lord," they respond, "We take refuge in God from Thee. Lo! we shall wait until our Lord cometh to us." God then asks them, "Is there a sign between you and Him by which ye will recognize Him?" "Yea," they say, upon which "God transforms Himself for them according to that sign"; thereupon, they say "Thou art our Lord. Glory be to our Lord." According to this highly symbolic imagery, in order to make Himself known to the believers who did not recognize Him in His unqualified state, God takes on an appearance that is recognizable to them, that is, in the form of the Hypostatic Face by which God had revealed Himself in each religion, for: "And for each [religion] is a [Divine] Face of which He is the Assumer. Compete, then, with each other in good deeds."[99]

Despite formal differences, all divinely instituted religions testify to the same universal truths. As unfamiliar as their symbolist languages may appear to today's mentality, they are highly consistent in their essence, that is, in their common universal principles, which stand in stark contrast to the implicit assumptions of modern science.

---

99 Koran 2:148. Let it be noted in passing that the fair competition, to which this verse calls religious adherents, could only be possible if all religions are equal in their truths and spiritual means, albeit, of course, in different divinely ordained ways.

CHAPTER 6

# THE INCOMPATIBILITY OF
# MODERN SCIENCE WITH RELIGION

The entire field of "science and religion" is dominated today by an ongoing debate over the reconcilability of the two. The representatives of the religions are often hard put to defend the reconcilability of religion with modern science in the face of vigorous assaults by the advocates of modern science, who have no shortage of evidence to point to the "unscientific" claims of the religions. Religious apologists labor hard to reconcile religion with modern science—often with little success, as witnessed by the ever-increasing number of people who renounce their faith in favor of science. Modern science, it seems, has become the criterion of truth: if the religions are not compatible with it, then they are of no account. In all this, we side with the advocates of modern science in their claim that modern science and religion are incompatible. We part ways, however, when it comes to what constitutes the criterion of truth.

In modern science, the perception of the knowing subject is erroneous because man's faculties of knowledge are reduced to his senses and reason, and that of the object of knowledge is likewise erroneous because reality is reduced to matter. Consequently, modern science is flawed both subjectively and objectively. It is flawed, therefore, in its very foundation, or rather in its lack of it. It is, as it were, a gigantic structure built on sand. The great religions, on the other hand, all benefit from integral epistemological and metaphysical doctrines, which base knowledge on the notion of the Absolute—hence on the notions of truth, objectivity, and certitude. Their perspective is thus solidly grounded.

By arbitrarily restricting the faculties of the subject and the scope of the object of knowledge, modern science condemns itself to fundamental errors. It is incapable of providing a sufficient view of the world, and yet lays claim to total reality. In Schuon's words:

> Those who seek to enclose the Universe within their shortsighted logic fail to be aware, at least in principle, that the sum of possible phenomenal knowledge is inexhaustible and that, consequently, present "scientific" information represents a naught beside our ignorance. . . . In all this wish to accumulate knowledge of relative things, the metaphysical dimension—which alone takes us out of the vicious circle of the phenomenal and the absurd—is expressly put aside; it is as if a man were endowed with all possible faculties of perception minus intelligence; or again, it is as if one believed that an animal endowed with sight were more capable than a blind man of understanding the mysteries of the world. The science of our time knows how to measure galaxies and split atoms, but it is incapable of the least investigation beyond the sensible world, so much so that outside its self-imposed but unrecognized limits it remains more ignorant than the most rudimentary magic. . . . A science, to truly deserve that name, owes us an explanation of a certain order of phenomena; now modern science, which claims to be all-embracing by the very fact that it recognizes nothing outside itself as valid, is unable to explain to us, for instance, what a sacred book is, or a saint or a miracle; it knows nothing of God, of the hereafter or of the Intellect and it cannot even tell us anything about phenomena such as premonition or telepathy; it does not know in virtue of what principle or possibility shamanistic procedures may cure illnesses or attract rain.

All its attempts at explanations regarding things of this order are vitiated basically through a defect of imagination. . . . One tries to explain "horizontally" that which is explainable only "in a vertical sense"; it is as though we were living in a glacial world where water was unknown and where only the Revelations mentioned it, whereas profane science would deny its existence. Such a science is assuredly cut to the measure of modern man who conceived it and who is at the same time its product; like him, it implicitly claims a sort of immunity or "extra-territoriality" in the face of the Absolute; and like him, this science finds itself cut off from any cosmic or eschatological context.[1]

Despite these fundamental deficiencies, modern man feels comfortable in his disjointed, contradiction-ridden world because:

Error creates the stage-setting it requires in order to feel comfortable. The world becomes increasingly a system of stage-settings destined to limit and distort the imaginative faculty, imposing upon it an unshakable conviction that all this is "reality" and that there is no other.[2]

Here a brief historical note may shed some light on the "stage-settings" modern science needed in order to establish itself as the authoritative account of "reality." Western science gradually diverged from the sapiential outlook that previously provided it with a metaphysical foundation, first by the as-

---

1 Frithjof Schuon, *Treasures of Buddhism* (Bloomington, IN: World Wisdom, 1993), pp. 41-44.

2 Frithjof Schuon, *The Transfiguration of Man* (Bloomington, IN: World Wisdom, 1995), p. 13.

cendancy of Aristotelianism especially during the late Middle Ages, and then by the subsequent renunciation of the metaphysical perspective altogether. Western science was thus cut loose from the higher authority of a metaphysical dimension. But an "extra-territorial" science, a physics without a metaphysics, for instance, is like a body without a head, or, to use a parable from the *Upanishads*, is like a chariot pulled hither and thither by the horses who hold sway while the charioteer has gone to sleep.

The pioneers of modern science made a conscious and deliberate choice to concentrate on the mechanism of things, and to ignore what seemed to them their troubling "why." In so doing, the materialist assumption of reality that underlay their empiricist project did not then seem to interfere with their genuine Christian beliefs since, as far as they were concerned, they had simply chosen a method—and no more— that would bring immediate and tangible results on the practical level. Only with the later identification of this method with science as such, would its materialist assumption of reality lay claim to total truth.[3]

In the absence of the intellectual backbone that is the sapiential outlook, the movements that were expected to meet the new sciences head-on—merely on the strength of their religious devotion and an assiduous study of the new sciences— eventually succumbed to the siren song of modern science— as do all such movements in the religious worlds today.[4]

---

3 Although the Scientific Revolution preceded the Age of Reason, it was not until the rationalist take-over of Western thought during, and after, the latter period that "thinkers" began to associate science as such with empirical science. It has to be emphasized, however, that the claim of empirical science to total truth is not "scientific"; rather, it is what may be termed "scientism"—something that superficially imitates true science—and which, often unconsciously, but sometimes deliberately, is taken to be "scientific."

4 The example of the Society of Jesus—an elite organization that originally rallied to the defense of the Church against new sciences—is not without relevance here.

The traditional sciences were generally crystallizations of universal principles—which are inherent in man's spirit and are awakened by the religions—and were thus organically related to the perspective of the great religions. Modern science—cut off as it is from such universal principles—often comes into conflict with religion. This conflict is inevitable given the divergence of the two outlooks.

Religious reformers have often tried to reconcile religion with science. Such reformers, in the worlds of Islam and Hinduism, for example, have sometimes wondered why the West—traditionally, Christendom—was the pioneer of modern science and technology whereas their worlds should have been a more favorable ground for the advancement of knowledge. They forget that it was not the West's attachment to Christianity that propelled it towards modern science and technology, but precisely its gradual abandonment of Christianity. Sacred sciences flourished in the worlds of Islam and Hinduism, for example, when these traditions were healthy. Nevertheless, the nature of their sacred sciences is diametrically opposed to that of modern science. One has the hereafter in view as its ultimate goal whereas the other is preoccupied with this world, and solely so. Wonders that the traditional sciences could work were no less impressive than those of modern technology, but their nature was not so much of this world. While modern technology has sent man to the moon, it was not uncommon for the saints of the traditional worlds to journey through much vaster worlds, of which our material world is but a small fragment.

There are some who wish to reconcile religion with science, not by revealing the symbolical significance that new discoveries may vehicle in their own way, but by interpreting religion in the light of scientific discoveries, as if the eternal truths of the religions were dependent on outward scientific data. What they forget, however, is that in doing so, in the final analysis, they are wishing to make the divinely revealed messages of religion subservient to modern science. Worse still, some would see the same goal for modern science and religion, and liken religion and spirituality to scientific exploration.[5] All this reveals a total ignorance of what truth is, namely, an immutable reality, which cannot be subject to any kind of becoming. It is the sign of a way of thinking that has lost its mooring, and is adrift in a sea of imponderables without issue.

In the face of the all-invading scientism of our time, and the collective psychoses resulting from it, the representatives of the religions, in nearly all contemporary religious spheres, find themselves pressured to draw parallels between modern science and religion. The evolutionist speculations that are so pervasive in the modern Catholic Church[6] provide a striking example of this kind of theology that has succumbed to scientism. In some Buddhist spheres, to give another example, it is nowadays taught that in cases of conflict with the findings of modern science, it is the Buddhist beliefs that should be abandoned. The divinely revealed religious dogmas, and the intellectually infallible doctrines deriving from them, are thus made subordinate to modern scientistic views, which are devoid of any intellectual foundation.

Another kind of attempt to reconcile religion with science comes from men of a scientific bent, who project their own scientific mentality onto sacred Scripture. The lack of a proper understanding of Scripture by these men, whose views are fundamentally shaped by modern science, paves the way for so many of the arbitrary pseudo-spiritual interpretations of the sacred texts in our time.[7]

---

5 Witness, for instance, the unfortunate positions of the fourteenth Dalai Lama in his *The Universe in a Single Atom: The Convergence of Science and Spirituality.*

6 Typified by the Darwinism of a Teilhard de Chardin.

7 A typical example may be found in Fritjof Capra's *Tao of Physics*, wherein the essentially materialistic and pseudo-spiritualistic interpretations, gained under the influence of psychedelics, of Lao Tzu's *Tao Te Ching* are

Thus, before any attempt is made to reconcile religion with science, it is important to understand the nature of modern science and its explicit as well as implicit assumptions, many of which are never seriously analyzed—in stark contrast to the nature and principles of the wisdom of the great religions, which are well-grounded in truth. In the ongoing debate about the conflict between science and religion, the advocates of modern science could have no better allies in the religious spheres than those apologists who have no real understanding of what modern science is, or, for that matter, what religions truly are.

presented in the name of science and the wisdom of Eastern traditions whereas, in reality, they have nothing to do with either modern science or traditional wisdom. Modern man's frenzied flight from a Law-giving Personal God sometimes leads him to take refuge in the Taoist and Buddhist traditions, of which he has little, if any, understanding, and which he interprets through the lens of his own ideological illusions.

# CONCLUDING REMARKS

## Modern Science and the Perennial Wisdom

By placing prime importance on man in his earthly contingency, the perspective of Renaissance and post-Renaissance man departs dramatically from that of traditional man—he, for whom the center, origin, and ultimate cause were always the Absolute, and in relation to which his identity was defined. The wisdom of the traditional worlds, appropriately called "perennial,"[1] and the science of Renaissance and post-Renaissance man are thus divergent in their principles, hence also in their outcomes, and their convergence is therefore impossible.

By ignoring epistemological foundations and metaphysical principles, modern science reduces man's intelligence to the lowest of his cognitive faculties, namely, his senses and reason, and reality to its most outward and contingent aspect, namely, matter. The perennial wisdom, or the *Sophia Perennis*, on the other hand, offers a full account of man's faculties of knowledge and a comprehensive description of the structure of reality. Traditional epistemology corrects the perception of the knowing subject, and the traditional doctrine of reality, or metaphysics, corrects that perception in regard to the object of science; these two are, therefore, the pillars of a true phi-

---

[1] The perennial wisdom is the timeless truth that underlies all the divinely instituted traditions, which are its providential vehicles. It finds its foremost expression in sacred Scripture, and has been taught and expounded throughout all the ages by great sages such as Hermes Trismegistus, Pythagoras, Plato, Plotinus, St. Augustine, Rumi, Meister Eckhart, and, down to our time, the preeminent exponent of the perennialist perspective Frithjof Schuon, from whom we have quoted extensively in this book.

losophy of science, in the absence of which no true knowledge can be gained. In Schuon's words:

> Modern science, which is rationalist as to its subject and materialist as to its object, can describe our situation physically and approximately, but it can tell us nothing about our extra-spatial situation in the total and real Universe. . . . Profane science, in seeking to pierce to its depths the mystery of the things that contain—space, time, matter, energy—forgets the mystery of the things that are contained: it tries to explain the quintessential properties of our bodies and the intimate functioning of our souls, but it does not know what existence and intelligence are; consequently, given its principles, it cannot be otherwise than ignorant of what man is.[2]

Unlike modern science, which perceives phenomena in their fragmentariness, traditional sciences normally started with principles, viewed their ramifications within the scope of their subject matter, and derived everything from them. The worldview implicit in the traditional sciences was "holistic," if one may use this term in this context, in the sense that the Universe was envisaged as a whole that essentially reflected the Principle, which is ultimately one, indivisible Reality. Things or beings were envisaged in their full reality by the traditional "scientist," whose more synthetic approach to phenomena may seem simplistic to the modern scientist, who dissects and analyzes.

The traditional sciences were often effective within their scope and goals. That they may not have been as successful as modern science in achieving certain practical and earthly goals is partly due to the fact that their goals too were "ho-

---

2 Frithjof Schuon, *Light on the Ancient Worlds* (Bloomington, IN: World Wisdom, 2006), p. 93.

listic," and had man's final ends in view—something which modern science ignores. And it is precisely this ignorance of man's final ends that constitutes the fundamental aberration of modern science when viewed from the integral perspective of the great traditions. In Schuon's words:

> We do not reproach modern science for being a fragmentary, analytical science, lacking in speculative, metaphysical, and cosmological elements or for arising from the residues or debris of ancient sciences; we reproach it for being subjectively and objectively a transgression and for leading subjectively and objectively to disequilibrium and so to disaster.[3]

Our criticism of modern science and our defense of the traditional perspective are not based on some nostalgia for the past. In this regard, to put things in perspective, let us say, along with Schuon:

> Inversely, we do not have for the traditional sciences an unmixed admiration; the ancients also had their scientific curiosity, they too operated by means of conjectures and, whatever their sense of metaphysical or mystical symbolism may have been, they were sometimes—indeed often—mistaken in fields in which they wished to acquire a knowledge, not of transcendent principles, but of physical facts. It is impossible to deny that on the level of phenomena, which nevertheless is an integral part of the natural sciences, to say the least, the ancients—or the Orientals—have had certain inadequate conceptions, or that their conclusions were often most naïve; we certainly do not reproach them for having believed that the earth is flat and that the sun and the firma-

3 Frithjof Schuon, *Esoterism as Principle and as Way*, p. 192.

ment revolve around it, since this appearance is natural and providential for man; but one can reproach them for certain false conclusions drawn from certain appearances, in the illusory belief that they were practicing, not symbolism and spiritual speculation, but phenomenal or indeed exact science. One cannot, when all is said and done, deny that the purpose of medicine is to cure, not to speculate, and that the ancients were ignorant of many things in this field in spite of their great knowledge in certain others; in saying this, we are far from contesting that traditional medicine had, and has, the immense advantage of a perspective which includes the whole man; that it was, and is, effective in cases in which modern medicine is impotent; that modern medicine contributes to the degeneration of the human species and to over-population; and that an absolute medicine is neither possible nor desirable, and this for obvious reasons. But let no one say that traditional medicine is superior purely on account of its cosmological speculations and in the absence of particular effective remedies, and that modern medicine, which has these remedies, is merely a pitiful residue because it is ignorant of these speculations; or that the doctors of the Renaissance, such as Paracelsus, were wrong to discover the anatomical and other errors of Greco-Arab medicine; or, in an entirely general way, that traditional sciences are marvelous in all respects and that modern sciences, chemistry for example, are no more than fragments and residues.[4]

Modern science is flawed in its principles and it cannot but be harmful in its results:

4 Ibid., pp. 192-193.

The sage sees causes in effects and effects in causes; he sees God in all things and all things in God. A science that penetrates the depths of the "infinitely great" and the "infinitely small" on the physical plane and yet denies other planes, even though it is they that reveal the sufficient reason of the nature we perceive and provide its key, is a greater evil than ignorance pure and simple; it is in fact a "counter-science," and its ultimate effects cannot but be deadly.[5]

The hypertrophied science of the modern era, then, is spiritually detrimental to man:

No piece of knowledge at the phenomenal level is bad in itself; but the important question is that of knowing, firstly, whether this knowledge is reconcilable with the ends of human intelligence, secondly, whether in the last analysis it is truly useful, and thirdly, whether man can support it spiritually; in fact, there is proof in plenty that man cannot support a body of knowledge which breaks a certain natural and providential equilibrium, and that the objective consequences of this knowledge correspond exactly to its subjective anomaly. Modern science could not have developed except as the result of a forgetting of God, and of our duties towards God and towards ourselves.[6]

### The Nature of Plenary Science

No truly integral and consistent philosophy of science is possible at the rational and phenomenal level, and if there can be a philosophy of science, it could only proceed from the knowl-

5 Frithjof Schuon, *Light on the Ancient Worlds*, p. 98.
6 Frithjof Schuon, *Esoterism as Principle and as Way*, p. 193.

edge of the one all-encompassing Reality, which comprehends all. This knowledge—the perennial wisdom—was the basis of all knowledge throughout all the ages up until recent history. Under its umbrella, the sciences at the periphery might once again be oriented towards a unifying and ordering center, which would then assure that they are indeed "reconcilable with the ends of human intelligence," that they are "truly useful," and that they are wholly "what man can support spiritually."

Reality, the ultimate object of true knowledge, is one; so is its ultimate subject. Total knowledge is situated beyond the bipolarity subject-object, because there is only one, indivisible Reality. The root of all knowledge is thus the identity between the knowing Subject and the known Object. Total knowledge means that the Absolute Knower knows Himself. In a sense, God is the Knowledge that the Divine Subject has of the Divine Object.[7] God can project this knowledge into man, and man is capable of participating in this knowledge because there is within him—or beyond him—a door that opens onto this knowledge. To actualize this knowledge, man has to transcend himself and become once again what he is in his essence. And that is precisely the goal of all authentic spirituality:

> Knowledge of the Total demands on man's part totality of knowing. It demands, beyond our thought, all our being, for thought is a part, not the whole; and this indicates the goal of all spiritual life. He who conceives the Absolute—or who believes in God—cannot stop short *de jure* at this knowledge,

7 We note that none of these statements constitutes a definition. The notions of "being," "science," and "happiness" remain beyond definition because in the final analysis they refer to the aspects of the Infinite—Being, Consciousness, and Beatitude, precisely—which cannot be contained within the bounds of any limiting definition; in Lao Tzu's words in the opening chapter of his *Tao Te Ching*, "The *Tao* that can be told is not the true *Tao*." Man has these eternal notions within himself because of the divine spark that is in him.

or at this belief, realized by thought alone; he must on the contrary integrate all that he is into his adherence to the Real, as is demanded precisely by Its absoluteness and infinitude.[8]

Metaphysical knowledge, to be actualized in man, requires of him all that he is. Now, man is essentially made of thought, will, and love:

> Thought of the true—or knowledge of the real—demands on the one hand willing of the good and on the other love of the beautiful, hence virtue, for virtue is none other than beauty of soul; that is why the Greeks, who were aesthetes as well as thinkers, included virtue within philosophy.[9]

Plenary knowledge engages all of man. It, therefore, consists of Truth, Way, and Virtue: metaphysical doctrine, spiritual method, and moral quality.

The starting point for true knowledge is thus a deep understanding of the evident Divine Unity, and its culmination is sanctity, that is, man's realization of the Divine Unity—the mystery of Union or Identity—which goes beyond the plane of the human mind. True knowledge is the knowledge of the Truth:[10] it is the identification of the subject with the object, which will then coincide with true happiness and everlasting freedom: "And ye shall know the truth, and the truth shall make you free."[11]

---

8 Frithjof Schuon, *Survey of Metaphysics and Esoterism*, pp. 5-6.

9 Ibid., p. 6.

10 "True knowledge is the knowledge of the Truth through the Truth, as the sun is known through the sun itself," says the ninth-century Sufi saint, Dhu'n-Nun al-Misri. And this is why the Sufi is said to be the "knower by God."

11 John 8:32.

# INDEX

# Index

# BIOGRAPHICAL NOTES

MAHMOUD BINA was born in Tehran, Iran, in 1938. He moved to England in 1958 to study engineering, but his quest for truth led him to pursue philosophy in Germany, where he received a doctorate from the University of Göttingen in 1969. Unsatisfied with the emptiness of modern philosophical thought, he came upon the writings of the perennialist author Frithjof Schuon, in whom he found his lifelong source of inspiration. He later moved to Switzerland and studied mathematics at the University of Lausanne, where he completed a master's degree program in 1973. After a teaching career at the University of Lausanne, he accepted a faculty position at Isfahan University of Technology, Iran, in 1977, a position that he held until his retirement in 2006. In addition to teaching a wide range of courses in mathematics, he taught philosophy of science for almost three decades, which left a lasting influence on countless students, for many of whom it marked the beginning of a lifelong journey in pursuit of truth. Dr. Bina and his wife live in Isfahan, Iran.

ALIREZA K. ZIARANI was born in Tehran, Iran, in 1972. He studied at Tehran Polytechnic where he completed an undergraduate degree program in electrical engineering in 1995. As a young man in search of truth, he was drawn to Professor Bina's philosophy of science classes, which exercised a profound and enduring influence on his life. In 1996, he moved to Canada to study mathematics at McGill University, and later engineering at the University of Toronto where he completed a Ph.D. degree program in electrical and computer engineering in 2002. His academic career brought him to the United States, where he was a tenured faculty member at

Clarkson University in Potsdam, New York, before he left for the private sector in 2009. As a professor, he taught courses in engineering and science, and led a biomedical engineering research program that was supported by the National Science Foundation and the National Institutes of Health. He currently resides with his wife in Bloomington, Indiana.